Introduction to Telephony

by

Jane Laino

THE TELEPHONY BOOK

Published by Telecom Books
An Imprint of CMP Media Inc.
12 West 21 Street
New York, NY 10010

ISBN 1-57820-035-0

For individual orders, and for information on special discounts for quantity orders, please contact:

Telecom Books
6600 Silacci Way
Gilroy, CA 95020
Tel: 800-LIBRARY or 408-848-3854
Fax: 408-848-5784
Email: telecom@rushorder.com

Distributed to the book trade in the U.S. and Canada by
Publishers Group West
1700 Fourth St., Berkeley, CA 94710

Manufactured in the United States of America

Introduction to Telephony is being used successfully by over 20,000 people who need a quick, basic understanding of the telephone systems and services used by today's organizations. The information presented is useful and easy to access. Author Jane Laino's expertise has developed from her early days working at the telephone company to her subsequent creation of DIgby4 Group, Inc., one of New York City's fastest growing telecommunications consulting businesses. Her down-to-earth approach translates confusing terms and complex concepts into plain English. This book is an important and unique resource for:

- ***Business People*** who purchase and manage telephone systems and services.
- ***Business Owners*** who want to understand their company's most critical system.
- ***Telecommunications Companies*** who need to train new staff members.
- ***Computer Telephony Developers*** who need to know what people expect from their telephone systems.
- ***Information Technology Professionals*** who are handed the responsibility for corporate telephone systems.
- ***Students and Educators*** who will supply the workforce with needed telephony expertise in the 21st Century.
- ***Investors in Telecommunications Companies*** who want to understand the industry and terminology.
- ***Bookstore Owners and Managers*** who offer rows of computer books for sale, but few basic books on telephone systems and services.

A quick flip through the Table of Contents and Index will preview what's in the book so you can see if it's right for you.

For tips on managing telephone systems, services and expenses
visit DIgby 4 Group, Inc.'s web site at www.digby4.com.

This book is dedicated to my parents,
Fern and Bill Coe

Thank you!

To Harry Newton, founder of Telecom Books and author of Newton's Telecom Dictionary. Harry developed the idea for this book in 1994. He realized the need for a basic book on telephony, particularly for those in the computer industries. The book was originally titled "Telephony for Computer Professionals." Harry sold the publishing rights to Miller Freeman last year, but he continues to be an inspiration to me and to many others in our industry. His enthusiasm and encouragement push us to new heights. His insights keep us moving in the right direction.

To Christine Kern, Manager of Telecom Books for her support, her creativity and her cheerful countenance. It's a pleasure to work with her.

To Steve Teta, president of American Data Transport, Inc. for his patient review of the book from a technical perspective.

To Gioia Ambrette, Kristi Mathus and all of the staff of Newcastle Communications in New York City for their help with the chapters on Voice Mail, Interactive Voice Response, Fax Server and Call Accounting. Their company specializes in selling and setting up these applications.

To Fran Blackburn of Intecom for her contribution to the chapter on Automatic Call Distributors. Intecom makes ACDs for Call Centers.

To Pattie Stone for her meticulous word processing services and creative suggestions. Pattie's support has been invaluable in getting this 3rd edition ready.

To the staff and associates of DIgby 4 Group, Inc. who keep our company growing and our clients happy – including Diane Ventimiglia, Elidé Chatley, Larry Napolitano, Bob Koehler, Debi King and Jerry Harder.

To the wonderful clients of DIgby 4 Group, Inc. for their confidence in us.

To all of my friends in the telecommunications industry each who has helped me in his own unique way.

– Jane Laino, New York City, 1999

About the Author...

Jane Laino spends her work days at DIgby 4 Group, Inc., one block south of Grand Central Station in the middle of New York City. She founded DIgby 4 in 1979, after 7 years with the Bell System (New York Telephone and Southwestern Bell) and 4 years with NatCom, Inc.

DIgby 4 Group offers professional assistance with planning, acquisition, implementation and management of telephone systems and services. What DIgby 4 really sells is peace of mind. Clients are comfortable that they are purchasing the appropriate telephone systems and services for the right price and that both the systems and expenses remain under control. DIgby 4 also consults in the new, but growing area of Computer Telephony – the addition of computer intelligence to telephone calls.

Jane lives in New York City during the week and spends the weekends on Candlewood Lake in Brookfield, Connecticut. She is married to Richard Laino.

She was born in Jersey City, N.J. and grew up on Long Island where she attended the Bayville Public Schools and Locust Valley High School. She is also a graduate of Queens College of the City University of New York with a B.A. degree.

Jane is a member of the Computer Telephony Pioneers and of the Society of Telecommunications Consultants. She also writes a monthly tutorial for TELECONNECT Magazine, a Miller Freeman Publication.

Her speaking, writing and consulting engagements keep DIgby 4 Group, Inc. up-to-date and Jane actively learning in the fast-moving telecommunications industry.

FOREWORD

Why I Asked Jane Laino To Write This Book

by Harry Newton

Many computer people are in for a nasty shock. They think because the technologies of computing and telecommunications are similar and the two are "merging," they'll feel right at home. They are wrong. Dead wrong. The phone business is completely different from the computer business. In every possible way.

How different they are is reflected in the fact that not one single computer company I can think of has made money in telecommunications. IBM lost over $2 billion. It bought Rolm, the PBX switch maker. It started Satellite Business Systems, the long distance phone company. And, at one stage, IBM even manufactured and sold its own PBXs in Europe. All these ventures are gone; closed or sold at a whopping loss. Ditto for Honeywell, which at one stage, owned a collection of interconnect companies – PBX telephone sales, installation and service companies.

The telephone industry has fared no better. AT&T lost so much in its own computer business (it sold its own Unix machines and private-labeled Olivetti MS-DOS PCs) that is was forced to buy NCR to cover its immense computer losses.

The local phone companies have fared even worse. When AT&T's Bell system was broken up in 1984 into AT&T and seven "Baby" Bells, the Judge made each of the Baby Bells "holding" companies, called RBOCs (for Regional Bell Operating Companies). This meant presumably, that they were meant to "hold" something. Like AT&T, they thought there was big money in the computer industry – presumably on the philosophy that things unknown are more intriguing that things we know about. Thus, in order to "hold" something, they bought everything in software and computer retailing they could get their hands on. Nynex even bought IBM's computer retail stores – the ones IBM couldn't make money on. At one stage, Nynex even tried to compete with IBM's money-losing Prodigy service. Nynex's venture lost oodles, also. Virtually every other RBOC got into software and lost more money than most of us ever dream about. One RBOC bought a software company for $340 million and sold it the following year for around $160 million. Today, all the RBOCs' computer dalliances are gone, at millions of dollars cost to their poor shareholders, who mistakenly thought they bought shares in a "safe utility."

To show how different the computing and telecom businesses are, let's look at each management discipline. Let's compare a computer company to a local telephone operating company, Bell or independent. Apologies in advance if this seems trivial. It's not. I don't have the space; but if you want to pursue this fascinating subject, e-mail me on HarryNewton@MCIMail.com.

Differences between a "computer" company and a local telephone company:

1. **Sales.** The computer industry is obsessed with selling. Every senior executive in any successful computer company is a great salesman. In contrast, telephone company executives are lousy salesmen. They've never had to be good – all their customers are captive. If you have a business in New York City, who else can you get dial tone from? What choice do you have?

   The telephone company actually has one "customer." It's called the Public Service Commission, the local state regulatory agency. That agency determines how much or how little money the phone company makes. The problem is that the local phone company doesn't see the agency as a "customer." It sees the agency as an adversary, an enemy who is always trying to reduce the telephone company's efforts and legitimate rewards. So the "sales" approach is always wrong. It's always, "Let me earn more money. Gimme. Gimme. Gimme." It should be, "I'll do this for you and the fine people of this State if you let me earn more money." The nice people who work at the PSC, who earn far less money than the telephone company executives who visit them, naturally resent this high-handed "Gimme. Gimme. Gimme." approach. And they typically do everything within their meager power to mess up the telephone executives' placid and pleasant lives.

2. **Innovation**. Computer companies only sell stuff if they sell new stuff. To live they must innovate. Phone companies do not innovate. They have essentially been selling the same service – 3 kilohertz switched phone service – for 120 years. There are a handful of new digital services which businesses, if you're lucky to be located in the right place at the right time, can get – T-1, digital Centrex, etc. But the list is miserably small. With only a handful of exceptions, the phone industry's new product marketing efforts have failed. Even touch-tone service, introduced into North America in 1963 (over 30 years ago), still has less than 70% penetration. In other words, over 30% of Americans have found touch-tone so unappealing they don't have it. The irony is that having its customers on touch-tone actually benefits the phone companies enormously because it lets them buy cheaper switching equipment and run their equipment fast and more economically. (They should be giving touch-tone away for free).

3. **Engineering**. There is innovation in the phone industry. It's just that it's internally focused. The phone industry has been quick to buy modern equipment which works better, more reliably, saves labor, etc. Testimony to this is that phone conversations today sound a lot better than they did ten years ago. And phone systems break a lot less often. As a result, the phone industry (AT&T, the Baby Bells and GTE) have fired over 250,000 people since divestiture, bringing down their costs. You and I, as customers, haven't seen those cost savings in our bills (with the single exception of long distance phone bills). Think of the cost/speed/performance improvement in your PC over the past ten years; think of the cost/speed/performance in your telephone bills over the same ten years. You should have had a commensurate improvement in telephone services. The two technologies are the same. Yet the industries are very different. What happened to all those freed-up monies in the phone industry? Try high salaries, waste, large expense accounts and all the millions lost in failed computer ventures.

4. **Executive motivation**. If you perform in the computer industry you do well. Your career prospers. Bill Gates says he only wants to hire people who've made serious mistakes. He won't hire anyone from the phone industry. No one in the phone industry makes decisions that fail. They just don't make decisions. The rule in the phone industry is simple: You'll never be penalized for not making a decision. So the tendency is never to make a decision. Go to meetings. Commission studies. Go to more meetings. Commission more studies. Don't ever do anything.

Here's a true story. New York Telephone employed me as a marketing consultant. I went to the first meeting. I asked, "Who's your largest prospect?" It was a customer who could give New York Telephone (now called Nynex) about $115 million a year in revenues and about $15 million to $20 million in profits a year. I asked, "Who's the salesman on the account?" Turns out that New York Telephone didn't have a salesman dedicated full-time to the account, and worse, the customer hadn't been visited in over four months. I suggested that they appoint someone. This recommendation took rocket science intellect on my part. Everyone on the third level management agreed. The vice president (fifth level) agreed this was a good idea. But the assistant vice president (fourth level) said that he wanted "to think about it" and have a meeting or two. I gave him till 5 PM the following day. At 5:05 PM, I walked into his office and asked what his decision was? He said he needed more time to think. I said there was no more time. I went upstairs to the vice president, who had already warned me that his $140,000 a year AVP was "useless." I told the VP that his AVP was being stupid and I recommended that he fire him on the spot. The VP answered, "You're 100% right. The man is stupid. He should be fired. But we can't do that around here." I immediately handed in my resignation. A couple of years later, I met the AVP. He had been given a promotion, a higher salary and moved to Nynex! I have no idea if they ever sold the account.

5. **Customer relations**. Computer companies want to please their customers. Telephone companies want to keep their customers from complaining. The BIG thing that gets a telephone

executive's attention is a complaint letter to the local PSC, or a complaint letter to the president of the telephone company. Computer companies want the excitement of new sales. Telephone companies want the peace of non-complaining customers who unquestioningly pay their monthly phone bills.

6. **Education**. *Go to any bookstore. You'll find acres of books on computers. Try to find the books on telephones.* This is not accidental. The telephone industry figured long ago that if it could keep its customers in the dark with a "Trust me, Trust me," philosophy, they could sell them and charge whatever they wanted. You don't believe me? Find your company's last month's phone bill. Ask yourself what all the line items mean? Ask yourself, "Are you really using what you're paying for?" Now, call your local phone company's business office and ask them to explain what's on the bill. Good luck. It's not because telephone services are complex. Computers are much more complex. It's because the telephone industry deliberately chooses to obfuscate how it bills for its products and services.

That's enough of the differences. Let's list a few of the more common mistakes computer companies make when dealing with phone companies:

A. **Motivation**. "If I do this," the computer company says, "your phone company will get $X thousand more per month. So let's do this joint deal together." Says the phone company, "Seems like a good idea to me. Let me study it." Translation: "When we've studied it to death in our various million committees, we might do a field trial. But the trial will be set up in such a way that it will fail. And then I won't ever have to make a decision to go ahead or not go ahead."

You, as an outsider, cannot assume that phone companies are motivated to increase their sales and their earnings. Remember, their earnings are limited by the local PSC. Whatever you do for them – and I stress whatever you do for them – you can never do what the local PSC can do much less painlessly and with much greater effect; i.e., get them money. Moreover, no phone company executive has any significant number of shares in his own company. Even at the highest level.

Virtually anyone that has ever relied on the phone company to do anything in a speedy, intelligent manner has gone broke waiting. I can present you with lists of companies. The acute lack of motivation translates into deliberate glacial speed, which will kill any outsider's motivation.

B. **Speed**. The phone company works at its pace, not yours. You may build yourself a palatial $100 million office complex. You want to be in on January 1. All your contractors may be on some form of motivation, bound by a written contract. Except one. The phone company. Your "written contract" with the phone company is the tariff which the phone company has filed at the PSC. That tariff typically states that if they're late installing their lines, tough. You wait. You can scream. And screaming often works. But it's very exhausting. In fact, dealing with the phone industry is always exhausting. In short, don't believe any promises they make about delivery. Ever.

C. **Homegrown.** The phone industry never hires from outside its own industry. That's not 100% true. They have been hiring

> some low level, not-too-bright executives. But you won't find any senior executives. Even IBM went outside for its new boss. You won't find that happening in the phone industry. The executives are very protective of their own secure, overpaid positions – after all, you, the customer, can still make a phone call, etc. So why should you complain? You sometimes sit in meetings with these people, as I have done, and wonder which planet they just stepped off.
>
> I think it's getting worse. To "fire" those 250,000 people, they offered many of them retirement and departure financial incentives that you'd have to be just plain stupid not to accept. You could argue, as a result, that the average IQ of the country's telephone company executives has dropped. You wouldn't be wrong. It's sad.

Is any of this going to change? The good news is that it will. It won't change because companies like Nynex change. They won't. They'll still be glacial, visionless, unmotivated and difficult to deal with. But they'll get competition. Technology is the great change merchant. It always has been in telecommunications. And it always will be. Technology brought MCI, Sprint and hundreds of others into long distance. That industry improved. Ditto for telephone equipment you put in your office. Ditto for fax machines, etc. Technology will bring competition to the local business. You'll buy local service from MCI, and dozens of others. They'll want your business and they'll do reasonable and wonderful things to get it. They'll cooperate with you in your computer telephony projects. They'll set up intelligent developer programs. They'll let you experiment. They'll let you put your equipment in their offices. They'll even take care of your equipment. Some of this might rub off onto some of the nation's phone companies. Bell Atlantic and BellSouth are acting more progressive. There is hope, yet. Just don't pin your business plan on it.

Harry Newton

New York City, June, 1994

Table of Contents

PART I – *The Basics*

PART II – *Hardware*

PART IV – *Interactive Systems*

PART V – *Computer Telephony*

PART I

The Basics

CHAPTER 1

Historical Perspective

"The Way It Was Affects The Way It Is Today"

Telecommunications Industry History: A Personal View

Many founders of today's telecommunications industry came out of the Bell System. The development of the industry has been shaped by their experience. The Bell System was essentially the telecommunications industry for many years. Bell rented you lines and equipment and sold local and long distance calls. Here's what it was like working for the Bell System from the late 1960's through the mid 1970's.

Some people worked in the craft or *plant department* end of the business and were members of the *Communications Workers of America* union (CWA). Installers, repairmen, field foremen, inside foremen, linemen; these were the workers on the front lines that were making it all happen. If you opened up a new business and ordered a telephone line and one telephone, the installer would show up on the due date, hook up the line, test it and install the telephone. It seemed deceptively simple. The installer usually had the order in his hand, written mostly in cryptic codes.

A copy of that same piece of paper was actually received and acted upon by forty different people within the telephone company. When you order a telephone line or even make a simple change in your telephone service today, the process is much the same. Many different departments have roles in the deployment of even the most basic services.

The *commercial department* included the business office. The people in this department were members of the *Union of Telephone Workers (UTW).* I started out at Forest Hills Commercial in Queens, a borough of New York City. I was a Service Representative. The business office was a structured environment, like the rest of the Bell System. Very little was left to chance.

Before starting to work, I was sent to a month long training class with a group of other women. The Bell System was big on training. The companies that used to be a part of it are still. We were trained exactly how to react and precisely what to say and do under every conceivable circumstance. If an inconceivable circumstance came up, we were to immediately put the caller on hold and run to get our "BOS" (Business Office Supervisor), the "Mother" of our unit. Each unit within the business office had approximately six representatives managed by a BOS. For every six units there were six BOS's and a District Manager to manage them. The District Managers reported to a hierarchy of other managers. There was a similar hierarchy in the Plant Department. The foremen were *first line, second line, third line,* etc., each line having a corresponding set of responsibilities, clout and compensation.

Back in the business office, our job was to take customer orders over the telephone. We released them to the order reviewers, who reviewed them. They sent them to the order writers, who typed them into the system for distribution to the other departments. It was also our responsibility to answer customers' questions about their telephone bills and, when we had time, to make collection calls to customers who were late in paying. These were known as "treatment calls."

Orders had to be written on a form in a very precise way, block printing and using the correct *USOCs (Universal Service Order Codes)* which were consistent throughout most of the Bell System. The terms *CV* for a single-line telephone and

KV for a six-button telephone sometimes heard today were USOC codes.

Another group supporting the business office was the *service observers.* The observers were hidden away in a nearby unmarked office. Throughout the day, they listened in on a random sampling of calls for each unit. This was to ensure that a high quality of service was maintained and that accurate information was being given out. The worst thing that could happen to a unit was to get a *scoring*, meaning that the service observer had heard something wrong. The observer would come racing into the business office and post the scoring on a bulletin board for all to see. Naturally, the members of the unit, including the BOS, were humiliated for the remainder of the day. We never knew which of the six members of the unit got the scoring.

There were different types of scorings. One was CWI (Customer Waiting Interval), meaning that a representative had left a customer on hold for an unacceptable amount of time. Another was ICR (Incorrect Rate), usually a result of someone adding the component rates of an order incorrectly (by hand – there were no adding machines on our desks). Another was IHM. I can't remember exactly what it stood for (Inhospitable Manner?) but it meant that a representative had handled the customer in a less than pleasant way. This was the worst type of scoring to get and one that promoted the most speculation about who did it and what was said.

When I transferred from New York Telephone to Southwestern Bell, first in St. Louis and then in Kansas City, the business offices were carbon copies of the one I left behind in New York. Nevertheless, each required a month in a training class before I was put back on the desk to talk to the customers.

The old Bell System was just that, a *system* and a very good one. The rigidity of structure and customer service monitoring worked well for a company which was regulated by the government and, as a monopoly, had no competition. As competition emerged, first for telephone systems, then for long distance service and finally, for local service, the Bell System's strong but inflexible structure became its weakness.

The layers of management made it difficult to reach a decision quickly. The uniformity of products and procedures throughout the United States prevented different parts of the organization from responding to conditions in the local marketplace. The tariffing of all services required that the Bell System obtain approval from each state's Public Utilities Commission prior to offering a new service. Brand new *Custom Calling Services* available from home telephones were introduced in 1969: Call Waiting, Call Forwarding, Speed Calling and Three Way Calling. More than thirty years later, the local telephone companies are still offering these same four services as "advanced" capabilities. And the business office is still pretty much the same as it was in the late sixties. The main difference is that customer records are now accessed by computer instead of paper files.

As monopolies, the Bell System telephone companies were regulated cost-based businesses. If they wanted to raise rates, they had to obtain approval from the government regulatory bodies and first had to make a case for why the rate increase was needed. The notion of making large profits did not exist, since it was not permitted. Thus, the Bell System was not the best training ground for the development of entrepreneurial skills. Many years after the AT&T divestiture of the Bell System companies and the deregulation of the telecommunications industry, some employees of the former Bell System companies still think of their business as it was. They have been challenged to adjust it to keep up with the changing marketplace and increased competition.

Back in the business office in the 1970's, if customers hinted that they were going to buy their own telephone system, we had a special "hot line" number to call to report this immediately. Someone from the telephone company sales department would then call to talk the customer out of making this big mistake connecting "foreign" equipment to the Bell System lines. I remember getting butterflies in my stomach just thinking that someone would be foolish enough to do this. That good old Bell System training included a fair amount of brainwashing as well.

The world of telecommunications began to change. Not everyone's entrepreneurial skills had been dulled by the Bell System. More business people were discovering the economic advantages of purchasing telephone systems rather than renting from Ma Bell.

In 1976, the *interconnect industry* was taking off. Interconnect companies sold, installed and maintained telephone systems and competed directly with the local telephone company who continued to rent systems. You could purchase a system for $100,000 that replaced the system you were renting from the telephone company for $10,000 per month. From a financial perspective, the 10-month break-even point was a no-brainer.

What businesses did not realize was that by buying their own telephone systems they were being pioneers. They often had a rocky road ahead.

It was around this time that I went to work for a consulting firm, back in New York City. Our specialty was showing businesses how to save a lot of money by purchasing their own telephone systems. We then proceeded to help them make the purchase and to manage the project. It was my job to manage the projects. My Bell System background was what got me the job, but it did little to prepare me for overseeing the installation of large business telephone systems. I started learning from the very first day when I was directed to show up at a *cutover* (this refers to the process of changing over from the old to the new telephone system).

I learned at that time that the local Bell telephone company was not very helpful to customers who had left the fold to buy their own systems. The local Bell viewed the people installing the new telephones as having taken away part of the Bell job, which they had, so Bell was not too anxious to make cutovers a big success. Sometimes the interconnect system installers had just quit their jobs at the telephone company and were now on the other side from their former co-workers. You still needed the local telephone company to install new outside lines or to make adjustments to the lines already in place to accommodate the new system.

At that time, organizations were also required to install *interface devices*. These were circuit boards of questionable usefulness connected to the end of each Bell System outside telephone line to protect it from the foreign equipment that was being installed. These interfaces were rented to the customer by the Bell System company. This was a consolation prize awarded to Bell by the courts in exchange for having given up the lucrative system rental revenue.

Actually, the interfaces just provided another point at which something could go wrong, which it often did.

The early systems did not always work as promised, so there were surprises. On several occasions, the installation company was unable to obtain the system that had been purchased so they put in a substitute system from another manufacturer, hoping that no one would notice.

I worked at a cutover where the customer had decided to purchase a used system. The installation company had not even bothered to remove the old coffee stains from the switchboard console! At this same cutover, it took a whole day before the installers could get the telephones to ring. When a call came into the switchboard, I would find out who it was for and run to tell that person to answer the telephone.

At another cutover, the customer had been told that the new telephones would be black, but the manufacturer only made white. That evening, the housings were removed from all the telephones and we spray-painted them black right there in the telephone equipment room.

At a typical cutover, a lot of people would be standing around looking worried, smoking cigarettes, drinking coffee and making periodic forays out of the telephone equipment room to reassure the customer: "Yes, at any moment now you will be back in business."

There was a lot of naiveté during this time. This included the buyers, the consultants, and the interconnect companies. We were all making our way through uncharted territory.

The telephone system functions were clunky and did not always work. To successfully transfer a call without cutting the caller off was a major accomplishment. Just to keep this in perspective, note that the first telephone system purchased by a company was often the first system that enabled the system users to do anything by themselves. It is likely that they had never dialed another extension or transferred a call before. The systems they replaced were often *cord boards* where the switchboard operator had been responsible for handling all of these functions for them.

Those were the seventies. Several decades have passed and the process of installing a new business telephone system has smoothed out considerably. At a recent cutover, our client announced that the change to the new telephone system had been a "non-event" and the business operations continued without missing a beat.

Purchasing Habits For Telephone Systems and Services

Telephone System Purchasing Habits

One thing hasn't changed much. Most organizations are still pretty much in the dark when it comes to purchasing telecommunications equipment and services. Almost every organization spends more than is necessary and often wastes considerable amounts of money.

Before the break-up of the Bell System, there wasn't any choice, so companies called their Bell System representative and bought what was available.

Suppose a business needs a new telephone system now. Here is how it often goes. Someone within the company is designated to be in charge of getting a new system. This person rarely turns out to be a hero.

The average business is dissatisfied with its present telephone system and with the company who sells and maintains it. The reason for changing telephone systems is usually one of the following:

— The company is moving.

— The company has outgrown the present telephone system (cannot add any more telephones or outside lines) or does not want to invest more money to expand an outdated system.

— The system has a lot of service problems which cannot be corrected.

— The system cannot accommodate new technology that the organization wishes to use.

There are two basic approaches to the new telephone system acquisition process. The first is to call in telephone system sales people and request proposals. The other is to develop a statement of your requirements and a Request for Proposal which you then submit to telephone system sales people so that you can compare proposals.

In Chapter 4 we talk more about how a telephone system is put together and how to buy one. As you will see, there are many variables in terms of hardware, software and capabilities.

Most companies do not put their requirements together first, other than the most basic information, such as how many telephones, how many outside lines and how much growth is needed. Then they end up with a stack of proposals that are impossible to compare since they are not based upon any common system configuration. As you might expect, the prices of these proposed systems vary widely. It is at this point that many people begin to realize that getting a new telephone system is not a simple process. Even with very small systems, a variety of options exist.

Usually a person responsible for choosing a new telephone system for a company decides upon one of the following alternatives:

— Selecting the system based upon which one has the "nicest" telephone instrument. The telephone sits on the desk and is the part of the system that everyone sees and uses. Thus, it is the telephone design itself that most often influences the purchase decision.

— Selecting the system based upon which sales person he/she likes best.

— Selecting the lowest cost proposal.

— Reporting to the boss that (a) the project is beyond his/her capabilities, or (b) he/she recommends bringing in some outside expertise to help with the process. This is not easy to ask since many bosses still view buying a telephone system to be almost as simple as buying paper clips.

In other words, most telephone systems installed today have not been selected through any methodical process. Some buyers even go to The Yellow Pages and call names that sound familiar.

The purchaser of a telephone system is usually one of the following people:

— In smaller companies, it may be the business owner.

— In medium-sized companies it is usually the office manager, sometimes teamed with the switchboard operator.

— As companies get larger, the responsibility may fall on a facilities manager, who is responsible for a variety of things including the office furniture, space planning, ladies room keys, office supplies and, oh yes, telecommunications.

— Businesses large enough to have an MIS department may have incorporated the responsibility for telecommunications into this department. MIS managers are focused on the computer network and tend to not pay too much attention to the telephone system. They are likely to hire outside support or delegate to in-house staff when faced with a telephone system replacement project.

— Still other businesses, usually only the largest or those spending the most money (such as Wall Street brokerage firms), have telecommunications managers and sometimes entire staffs responsible for selecting, implementing and managing telecommunications systems. These companies usually have an MIS manager as well. Communication between the telecommunications manager and MIS manager is often minimal.

— Some firms retain the services of telecommunications consultants to help them to assess their requirements and select a system. The consultant usually makes a recommendation, but is not always the final decision-maker on which system is selected. The consultant does, however, put together the statement of the system requirements on which the proposal is based.

Some telephone system contracts are very detailed, showing each system component. Others are very general. We have even seen some that do not provide the name of the system manufacturer. It is better to get as much detail as possible, provided that it is clear what the details mean.

Chapter 3 provides more information on the types of companies from whom you may purchase a telephone system.

Common Mistakes in Purchasing Telephone Systems

Here are some of the most common mistakes companies make when purchasing a telephone system:

— Buying a system that cannot accommodate growth. Even if you do not plan to add many telephones, you may want to incorporate other capabilities into your system which require expansion space in terms of both hardware and software.

— Buying a system that is at the end of its life cycle. Systems are sold right up until the day they are replaced by a newer model or a new software release.

— Buying a system which is not installed at many other businesses in your area. If the company who installs it for you changes its product line or goes out of business, you may have trouble getting maintenance support.

— Buying a system whose functions do not complement your organization's operations.

— Buying a system that will not enable you to accommodate new technology and your organization's changing requirements.

Tips For Purchasing Telephone Services

The people who are responsible for the telephone system are usually the same ones responsible for purchasing complementary telephone services. This includes outside lines and local and long distance calling services.

Purchasing long distance calling services, like purchasing telephone systems, is not as simple as it might seem. Although the local telephone companies may be selling long distance calls, there are still basically two other types of companies from whom you can purchase long distance service. The first type is the long distance carrier. The big three are AT&T, MCI/Worldcom and Sprint, but there are other smaller carriers. The second option is to buy from a company who is in the business of reselling long distance service which they purchase from one or several carriers. There are numerous types of resellers, providing service in many permutations. One of the challenges that each of the carriers and resellers faces (since there are so many of them) is how to distinguish oneself from the pack.

These long distance carriers and resellers also sell point-to-point dedicated lines connecting, for example, your New York office to your Chicago office.

Most companies select long distance service with cost in mind. All other things seeming equal, the objective of reducing and controlling expenses prevails in the decision-making process. In most organizations, the largest chunk of the telecommunications expense is for long distance calling, far outweighing the cost of equipment and outside lines.

Due to many options for service and many components determining long distance calling rates, it can be difficult to monitor costs and to verify that predicted cost reductions are realized. Here are some of the variables:

- **The cost per minute**. When trying to track long distance costs and savings, it is a good idea to come up with a cost per minute or average cost per minute for different types of calls (intra-state, interstate and international). Find a fixed cost per minute regardless of where you call. It will make tracking expenses a lot easier.
- **The number of minutes**. This can vary depending upon both the actual time spent on the calls and the type of rounding done on each type of call. For example, some carriers round each call in six second increments, so if you were on a call for thirteen seconds it would be rounded up to 18 seconds. Some bill in one second increments. Some older long distance services still in effect may round to 30 second or even 1 minute increments. This can add substantially to the expense.
- **The type of service for which you are signing up**. Most long distance companies have a variety of plans, with

marketing names that change regularly. Some service is *equal access* which means that the long distance calls go out over the lines coming into your premises from the local telephone company. [*Note*: Most long distance companies now sell local service competing with the traditional "local telephone company."] Other services connect you directly to the long distance carrier's switch with a separate high-capacity line, such as a *T-1* (see Chapter 9).

- **The length of the contract.** Usually, if you sign up for a longer period of time you will pay a lower cost per minute. If the carrier lowers its rates while you're still in a long-term contract, you may ultimately wind up paying more, so you may not want to lock yourself in for too long a period. Be sure you know what termination penalties exist if you need to get out of the contract.
- **The quality of customer service**. If you have a question on your bill, it's important that you have a place to call where you can get a fast and accurate response. You want the same if you are having problems with your long distance service. It's also important to have an account executive who will visit you regularly and inform you of new and better services as they are introduced. Put this into your agreement before signing up. None of this can be taken for granted. Find out what you can expect before starting a relationship with a long distance company.
- **The format of the bill.** In terms of tracking costs and verifying projected savings, an accurate, easy-to-understand bill providing a lot of detail will make the task easier. Bills may be sent in a paper form, on disk or CD-ROM. Request copies of the format of bills you will receive. Find

out if customized reports are available. Some carriers also enable you to access your call details at their web site, even before the month is up.

Purchasing local telephone service is yet another area in which there are many variables to consider. These include the type of line, cost and whether or not the cost will be lower if you agree to sign up for a service for a longer period of time. Some contracts go up to 120 months (ten years).

Chapter 9 provides a description of the different types of local circuits and how they are used. You may rent these circuits from your local telephone company. In most areas, the local telephone company has designated a group of other companies as its agent. Agents may be companies who install and maintain telephone systems.

As mentioned, the traditional local telephone companies now have competition for local circuits and services. This is starting in cities and other areas where there is a heavy concentration of business customers. The local cable TV companies are also providing local telephone service in some areas such as New York City.

When seeking advice, it is important to know that most people working in the telecommunications area know a lot about some things and very little about others. Some are very experienced in the systems area but don't know the first thing about looking at a telephone bill. Others can design a sophisticated high speed network for voice and data communications, but don't know how to transfer a call on a PBX. So choose your sources of expertise and support carefully. Question everything!

Each chapter of this book will give you more points to consider and questions to ask when purchasing telecommunications equipment and services.

Telephony vs. Computer Industry Cultures: Why Can't We All Just Get Along?

As Computer Telephony integration becomes a reality, the cultural differences between the telecommunications and computer industries are becoming more apparent. Hopefully, we will all be enlightened enough to draw on the strengths of each and learn to tolerate the inevitable differences of opinion.

One critical area where the telecommunications types can teach something to the computer people is that of record keeping. As companies invest more money in equipment and services, it becomes more important to keep track of costs. Most companies know what type of telephone is on each desk and what that telephone can do, but those same companies seldom track PCs and the software in use on each of them. The need for cable records and circuit records are also common to both telephone systems and PC networks.

Many large organizations maintain separate telemanagement systems not only for tracking their telephone system and telephone call records, but for the purpose of charging back the costs of these to the departments. Traditionally, corporate management has not placed this same requirement on computer systems since they have been more centralized. As PCs on the desk and Local Area Networks become the norm, these same types of controls will need to be put into place. The dramatically increasing use of the Internet by employees has raised a flag to corporate managers everywhere. They realize they are losing control over workforce activities. This requires putting recordkeeping and control systems in place.

Another strength of the telecommunications industry is its experience with project management. Much of the orderliness and structure stem from the Bell System traditions described earlier. Experience with training and putting together training materials is yet another telecommunications industry tradition.

How about standards for reliability? Everyone expects the LAN to crash every now and then, but let just one telephone be out of service for a few minutes...

The telecommunications industry has much to learn from the computer industry as well. We mentioned earlier that the "advanced" features offered by the local telephone company today are the same ones that were considered to be "advanced" in 1969. Meanwhile, we are receiving upgraded versions of software for PC programs every few months which incorporate an array of new features.

The computer industry comes out with more new and innovative products in a month than the telecommunications industry does in more than a decade. If the leaders of the telecommunications industry don't speed up the rate of change, they will be left in the dust or will be merely providers of commodity circuits requiring computer software to make them function. Many people already view telecommunications circuits as merely the bricks and mortar to be used and controlled by software as parts of a network for voice, data and image or video communications.

People in the computer industries tend to be analytical and innovative when approaching problems and designing systems. They are less bound by traditions and conventions than the people who have been designing telephone systems.

Having worked with telephone systems and services for many years, I can tell you that the way things operate does not always make a lot of sense. Systems have not been designed with a lot of thought as to how people work. Companies are bending the way they work to fit the constraints of their telephone systems.

My view of computer/telephony integration is that it is an opportunity to wipe the slate clean on poor telephone system design. It would be a shame if the result were just the same old cumbersome features appearing on the computer screen instead of on a separate telephone instrument.

As someone who grew up in the telecommunications industry and has now entered the world of Computer Telephony, I am hoping for a cooperative spirit between the two industries.

A View From DIgby 4 Group, Inc.

As president of DIgby 4 Group, Inc., I observe the telecommunications industry from many different perspectives. To me, the most important view is that of the customers who purchase telecommunications equipment and services.

As the industry becomes more complex, customer decisions become harder. As consultants who provide both expertise and assistance in making things happen, we receive many calls for help. Here are some examples of why people call us:

We Can't Get New Outside Lines Installed

Q. *We ordered two new outside lines from our local telephone company two months ago. They still can't tell us when they will be installed.*

A. The problem here is that it was going to cost the local telephone company $20,000 worth of labor and materials to run new cable since all cable into this building was already in use. They were understandably dragging their feet. The customer only pays $150 for the installation of the lines and $30 per month each. Even with the revenue from local calls, it may be a long time before the telephone company recovers their investment. If the customer changes to a competitive local service provider, the investment may never be recovered. In this case, once the customer agreed to a 5-year contract, the company ran the cable.

We've Invested Money In New Equipment And Callers Are Still Complaining

Q. *We recently invested over $30,000 in an Automated Attendant and Voice Mail system to answer the calls to our customer service center. Callers are complaining that they are on hold forever. Why is this happening?*

A. First, the new automated system was installed to work with a very old telephone system. The two systems did not communicate very well. Second, since the automated system was purchased from a different company than the one who services the telephone system, the phone system vendor did not have much of a stake in making it work. Third, although an excellent automated system was purchased, it was not the right type of system for the circumstances. The client now has an Automatic Call Distribution system and callers have ceased to complain.

Will It Really Cost Us $600,000 To Add Five Telephones?

Q. *We need to add five more telephones to our system. We have been told that since our system is at maximum capacity and is no longer manufactured, we must buy a new system. With 500 telephones, this will cost over $600,000. There is no money in the budget. Is there any way around this?*

A. There are still a lot of telephone systems out there that were installed more than 10 years ago. Some of them are large systems and still work very well for the way they are being used. This particular one was in a non-profit organization. As it turned out, the system could be upgraded to a higher release of software that cost about $20,000. This enabled growth of 100 more telephones. Finding this out was not easy since the equipment vendor would have much preferred to sell a new system.

Will Our Callers Ever Be Sent To The Right Department On The First Try?

Q. *We had six switchboard operators. Each of them had been with our company for over ten years. We recently outsourced the entire switchboard answering operation. Now callers are complaining that their calls are ending up in the wrong departments.*

A. We suspected that similar problems occurred with the former operators. (These operators were required to train the new ones that replaced them!) Our suggestions here focused on two areas: (1) Having the individual departments improve their capability to handle their own directly dialed calls and (2) setting up a directory system so the switchboard operators can do a better job of sending callers to the right place on the first try. Getting callers to the right person and keeping directories current is a problem common to most organizations.

Must We Give Up Our Centrex System To Keep Control Of Our System Program Changes?

Q. *Our organization is new and growing rapidly. We are using Centrex service and have the ability to dial into it to make program changes to our telephones. Now new telephone instruments have been introduced and our president wants to use them. But the local telephone company does not yet have* *the capability for us to do our own programming on these new telephones. Instead we must call them and wait for days to have the changes made.*

A. This organization had assessed the situation correctly. We had no suggestions for changing this. Right now they are living with the older style telephones. We dial in from our offices and make the program changes for them within several hours of their request (changing line appearances on telephones, adding intercoms, etc.) The next step will be to compare the cost of keeping Centrex to buying a new PBX that will give the company more control of their telephones, but also the responsibility of maintaining and managing an on-site telephone system.

We're Changing The Business We're In – So We Need A New Telephone System

Q. *Our family has been in the answering service business for three generations. Now that Voice Mail has taken over, few people are using live answering services anymore. We have gone into the new business of taking orders over the telephone for other companies. The problem is that we are using the telephone system we had used for the answering service business and it does not give us much flexibility. We want screens to be customized so our representatives can have different scripts and different screens for taking orders for our various customers.*

A. This was the perfect opportunity to suggest Computer Telephony. The new system turned out to be a marriage of a PBX, a LAN and software designed for incoming call centers. The cost was $150,000 for 14 customer service representatives. Much of our time was spent getting the telephone system vendor to talk to the computer installation company and then getting both of them to talk to the software company. Now, when a call comes in, the appropriate screen pops up. The representative can then take an order for whichever company was called. In addition, our client's customers can dial in from their PCs to see how many orders have been taken. One irony is that the system that was replaced, designed specifically for answering services, was a true computer telephony system with no telephones. The headsets connected to the terminals. The new system requires a separate telephone and a PC on each representative's desk. (*Note*: This was completed in 1994.)

Computer Professional Hates Phones

Q. *I run a computer software company and I hate telephones. My company is moving and we need a new telephone system. I want to experiment with computer telephony so that my smart little computers will tell that big dumb phone system what to do. Can you help me?*

A. This client was a challenge, but we proceeded to bring him to see demonstrations of six different telephone systems with varying capabilities to interface with computers. The client grilled the telephone system vendors mercilessly on how their systems worked in the computer telephony environment. Most did not know the answers to his questions which frustrated him to no end and reinforced his disdain for telephony. Nevertheless, he finally did settle on a system which offered a reasonably priced and flexible capability to interface with the LAN. To save time and money, this same client decided to reprogram his six-year-old Voice Mail system. Despite our reservations on this do-it-yourself approach, he proceeded to "crack the code" and got the Voice Mail working with his new telephone system.

Long Distance For Less?

Q. *Last year we installed a T-1 circuit to our long distance company. We were supposed to save money. Our bills seem to have gone up instead of going down. We are not sure what is happening. Can you take a look at this for us?*

A. What was happening is that the T-1 was sitting there in the client's telephone system, but the telephone system vendor had never programmed it into the PBX's Automatic Route Selection. In short, no calls had ever gone over the T-1 at the reduced rate. Although a year had passed, the long distance carrier had never brought this to the client's attention. All long distance calls were being sent over the regular outside lines using a different long distance carrier who was billing at the maximum rate. (14 cents a minute as opposed to 5 cents a minute if calls had gone out on the T-1.) To make matters worse, the long distance company providing the T-1 was now charging the client $42,000 for "underutilization." We were able to have that charge waived. The telephone system has now been reprogrammed to send calls out on the T-1.

Buy Experience And Support

Q. *We are a large university. Our Voice Mail system is awful. It cuts off callers and gives out inappropriate recorded announcements. Our staff has difficulty in retrieving their messages. Sometimes messages are lost in the system and reappear weeks later. Everyone has lost confidence in this* *system. We just bought it two years ago. Can it be improved or must we replace it?*

A. What happened here is that our client had purchased the system inexpensively from a gentleman who put the system together himself, buying the PC and voice boards and writing the program. This was his first and last foray into selling and installing Voice Mail systems. He is now in a different business. There was no documentation in terms of how the system was programmed and our client did not have access to the source code. Our recommendation was that they buy a new system from a company who had experience and a support organization behind them.

Don't Wait To Think About The Phones

Q. We are moving to new offices. We will be purchasing a new telephone system and will need some help with that, but right now our architect is finalizing the floor plans. She is asking us questions we cannot answer about what will go into the telephone equipment room. Can you send someone over for a meeting this afternoon?

A. We often find ourselves meeting with architects and engineers who are designing a space. They discover that their clients have not yet begun to think about the telephone or computer systems that will be installed in that space. We typically provide them with estimates in terms of requirements for space, electricity and air conditioning. Unfortunately, once the client actually decides which telephone system to buy and what the computer requirements will be, this information often changes. The point is that the planning and selection of technology systems should take place as early as possible in the designing of a new space.

An exception to this is a client who has retained us, but is not moving for two more years. He wants to wait to see if technology changes within the next year before selecting new systems.

You're Paying How Much?

Q. *We think we are doing a good job of managing our telecommunications expenses, but our boss read this magazine article on the plane. Now he wants someone to come in and take a look at the telephone bills. Is this something you can do?*

A. Most organization's telecommunications expenses are higher than they need to be. Sometimes we find overbilling and are able to obtain substantial refunds ($100,000 or more). At other times, no refunds are due, but ongoing costs may be dramatically reduced by up to 50%. Sometimes circuits are disconnected, but the billing continues. In other cases, services are billed at an incorrect rate. We also see great discounts negotiated for long distance calls, but in looking at the bills, find that the discounts have not been applied! We see companies paying $75,000 for a telephone system that should have cost $40,000. We see an incredible amount of waste. Question everything and shop around!

These are just some examples of what purchasers of telecommunications equipment and services encounter every day.

CHAPTER 2
The Telephone

Newton's Telecom Dictionary defines the telephone as a truly remarkable invention that does the following things:

- When you lift the receiver, it signals the local system that you wish to use the worldwide phone system.
- It indicates that the phone system is ready for you to dial by giving you a dial tone.
- It sends the number of the telephone to be called.
- It indicates the progress of your call through tones: ringing, busy, etc.
- It rings to alert you to an incoming call.
- It transforms your speech into electrical signals for transmission to a distant point and translates the electrical signals it receives back into the human voice for you to hear.
- It automatically adjusts for changes in the power supplied to it.
- When you hang up, it signals the telephone system that you are finished.

The telephone is also referred to as a telephone instrument, *station* or set. There are *single-line telephones* and *multi-line telephones*. These telephones can be analog or digital and you cannot always tell which is which by appearance. Telephones work with outside telephone lines from a local telephone company or as extensions from a business telephone system.

The telephone is designed to operate under a wide range of electrical, mechanical and acoustical conditions. Some of the design parameters are dictated by human factors such as sound pressure levels and handset dimensions. Some are historical carryovers such as ringing voltage and frequency. Others, such as the minimum line current for satisfactory carbon transmitter and relay operation, are dictated by the physical properties of the materials used in the telephone.

Telephones in use today are of different vintages. The more recently manufactured telephones substitute microphones for some of the materials such as carbon transmitters still found in some older telephones in use.

The following pictures show a single-line telephone (Figure 2.1) and a multi-line telephone (Figure 2.2). As you can see, they vary somewhat in appearance, but have some things in common.

Figure 2.1
Traditional Single Line Telephone
(Also called 2500 Set)

Courtesy of CorelDraw

Figure 2.2
Multi Line Telephone

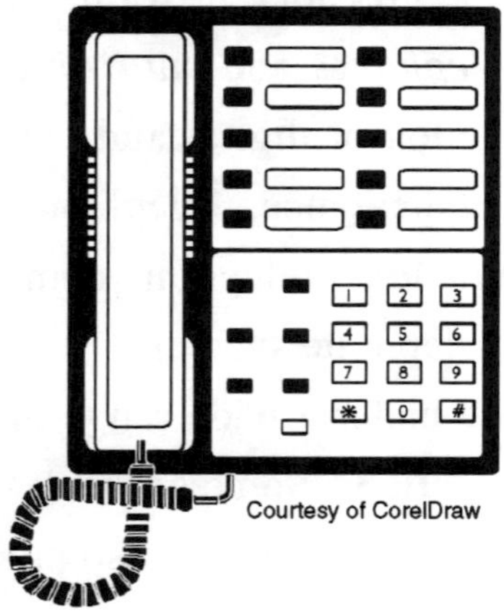

Courtesy of CorelDraw

Parts Of The Telephone

First, we will talk about the things you can see, and next we will get to what is inside. Here is a telephone with its external parts labeled.

Figure 2.3

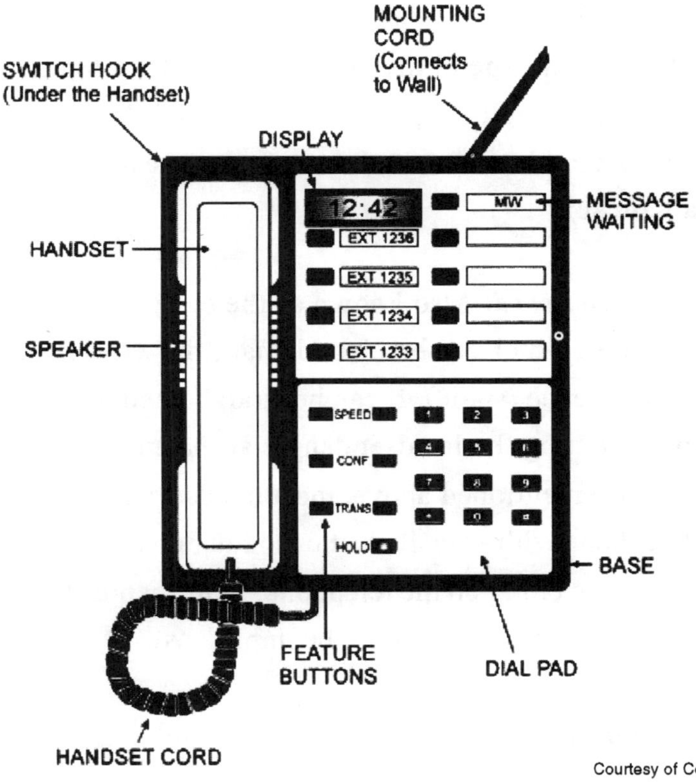

Courtesy of CorelDraw

- **The *telephone handset*, also called the *receiver*.** In fact, it includes both the receiver enabling you to hear and the transmitter through which you speak. It may also have a volume control or a bar that you can depress to mute either the receiving or transmitting capability.

  Handsets come in different shapes and sizes and are often made to work with a telephone from a specific manufacturer.

The handset may be hardwired to the telephone cord which in turn is hardwired into the telephone instrument, or there may be a modular connector at one or both ends of the cord.

It is possible to buy a handset separately from a telephone and plug it into some type of jack, so long as that jack is wired to the telephone.

People who spend the entire day on the telephone such as customer service representatives, stock brokerage traders and switchboard attendants often use a *headset* instead of a handset.

- **The *handset cord.*** Also known as the *curly cord.* Often gets very twisted which can break or damage the wires inside causing interference (static). It can be straightened out by holding it up and letting the handset dangle at the end, unwinding itself. As we mentioned above, most handsets are connected to the telephone with a small plastic modular connector that plugs into a jack opening on the telephone. Some handsets are wired into the telephone and cannot be unplugged.

- **The *mounting cord.*** A straight cord or cable, usually gray or silver. Typical lengths are 6 feet, 9 feet, 13 feet and 25 feet. This cord sometimes has a modular connector at both ends, one plugging into a jack opening on the telephone and the other plugging into a jack opening in the wall. In some cases the mounting cord is wired directly to the telephone or the wall or both and cannot be unplugged.

- **The *dial pad.*** Also called the *keypad, touch-tone pad, touch-tone buttons* or *DTMF pad* ("DTMF" standing for dual tone multi-frequency, referring to the touch-tone signals). Most telephones use the DTMF method for sending a telephone number.

The telephone company central office must have the capability to process these tones. The telephone is equipped with the dial pad having 12 buttons that represent the numbers 0 through 9 and the symbols * and #. Pressing one of the buttons causes an electronic circuit to generate two tones. There is a low-frequency tone for each row and a high-frequency tone for each column. Pressing button number 5, for example, generates a 770-Hz tone and a 1,336-Hz tone. By using this dual tone method, only seven tones produce 12 unique combinations. The frequencies and the dial pad layout have been internationally standardized, but the tolerances for variations in frequencies may vary in different countries. If you plan to have callers from other countries use your automated telephone systems, be sure the systems can recognize the touchtone signals.

The dial pad is used not only to dial telephone numbers, but to interact with Voice Mail and interactive voice response systems. (For example, Dial 1 for Sales, Dial 2 for Accounting, etc.)

Some telephones may have a rotary dial, but this is becoming less common. The signals sent out by a rotary dial are called *dial pulse.*

There are still many rotary dial telephones in use. In general, their dial pulses are not recognized by voice processing equipment. Some of the newer voice processing systems get around this using *voice recognition* (Dial 1 or say "Yes" for the Customer Service Department.)

Some telephones have a little switch on the side which converts the signals being emitted by the telephone from

DTMF to dial pulse. This will be necessary for placing a call from a touchtone telephone if the local telephone company central office cannot accept DTMF signals. The dial pad is still touch-tone, not rotary. When you punch in the called telephone number on the touch-tone pad, it will take longer for the telephone to send out the dial pulses (about as long as it would take if you were actually using a rotary telephone). You can usually hear the sounds of the telephone "pulsing out" while you are waiting.

On occasion, you may encounter a telephone working with a business telephone system from which no DTMF signals are sent. This presents a problem when trying to use the telephone with voice processing systems.

Some telephone systems do not emit a DTMF signal when the "#" (pound) or the "*" (star) button is pressed. There is usually a fix of some type that will correct this.

- **The *feature buttons*, also known as *feature keys* or *function keys*.** These can serve a variety of functions. They enable different outside lines and extensions to be answered. They may activate telephone system functions such as call transfer, call conferencing, call forwarding, etc. They can also be used to speed dial frequently called numbers. Every telephone system manufacturer treats these feature buttons differently, so **what you learn about one system may not apply to another.** Some feature buttons are flexible, meaning that they can be programmed for a variety of functions. Some are fixed, meaning they can provide only one function. Some systems have *soft keys* meaning that the same button performs different functions at different times depending upon what is displayed on the corresponding LCD. Later in this chapter we will review some of the features and functions that can be performed by the telephone.

- **The *display*, also known as the *LCD* (liquid crystal display).** Not all telephones have displays. Most system manufacturers provide them, but the display telephones cost more. Different systems provide different information in the display. Some show the date and time when the telephone is not in use. Others provide instruction prompts to the person attempting to use telephone system functions. Most show the name or extension number of the person calling you, if the call is coming from someone else within your office. Some show the name or telephone number of the person calling you from another location. Some systems enable you to leave a preselected message so that when someone calls your telephone from within your office, his display will read that you are "out to lunch" or "in a meeting." Other systems enable a secretary to send a silent message to the boss while the boss is on another call. As with the feature buttons, the important thing to remember is that telephones and telephone systems from different manufacturers use the display differently. No two are exactly the same.

- ***Lamps*; *lights*; *LED* (light emitting diode).** The purpose of this is to indicate the status of a call in progress on one of the outside lines or extensions. The lights may be red, green, white or amber. This differs depending upon the manufacturer and system. A lamp flashing on and off slowly may indicate a new incoming call and is sometimes accompanied by an audible ring. If the same extension appears on more than one telephone it may simply flash at some telephones and audibly ring on others. A steady lamp usually indicates that the line is in use on either your telephone or another telephone that picks up the same line. A rhythmically flickering lamp may indicate that a call is on hold. On some systems there may not be a lamp, but an LCD indicator instead.

- **The *switchhook.*** This refers to those two little plastic buttons that press down on a traditional telephone when you hang up the receiver. When you hang up you are actually breaking an electrical circuit which connected you to the person at the other end while you were talking. On some telephones, it may be a single bar that depresses when you hang up. Other telephones have a magnetic switchhook that cannot be seen from the outside of the telephone. In old movies, we see someone frantically tapping the switchhook trying to get help as the intruder is banging on the front door. This method was once used to reach an operator or switchboard attendant. You would now dial 0

Figure 2.4

Courtesy of CorelDraw

to accomplish this. In telephone systems introduced in the 1960's and 1970's, the switchhook was used as a means of activating the system functions such as call transfer. These are now performed more easily with feature buttons.

- ***Speaker.*** Most multi-line and a few single-line telephones are equipped with some type of speaker. A *speakerphone* enables the person using the telephone to have hands-free conversation with another person at a distant location without lifting the handset. Some speakers are one way only. These enable the person using the telephone to dial out or wait on hold. They can hear what is on the open line, but cannot speak back to the caller without picking up the handset. A speaker in a telephone may also be used for internal intercom communication only, where someone on the same premises can call you and the voice will come out of the speaker. Some systems will allow you to answer back hands-free while others will not.

- ***Message waiting indicator.*** If the system is working with a Voice Mail system, this lamp or LCD indicator lets you know that you have a message waiting in your Voice Mailbox. It may also indicate a message waiting at the reception desk or message desk if there is no Voice Mail, although this use is less common. In some systems, the message waiting indicator is a button that, when depressed, will send you right into the Voice Mail system to retrieve your messages.

- **Base of the telephone; telephone housing.** This is generally a molded plastic casing designed to house a specific type of telephone.

Now, let's get to what is inside the telephone.

- **The *transmitter*.** (Figure 2.5) The transmitter is the ear of the telephone in that it "hears" the voice of the person speaking into it. The transmitter is a miniature carbon pile rheostat. A *rheostat* is a device that controls an electric current by varying the resistance in the circuit, similar to the action of a dimmer switch control. The variations in sound pressure from the voice vibrating against the diaphragm change the compression of the carbon granules. This varies the resistance of the transmitter. The transmitter has two contacts that are insulated from each other. Current can only flow through the carbon granules. As sound pressure from the voice presses against the diaphragm, the carbon is more closely compressed within the chamber. Compressing the carbon granules lowers the resistance of the transmitter resulting in more current flow through the transmitter circuit. When the pressure on the

Figure 2.5

The Transmitter

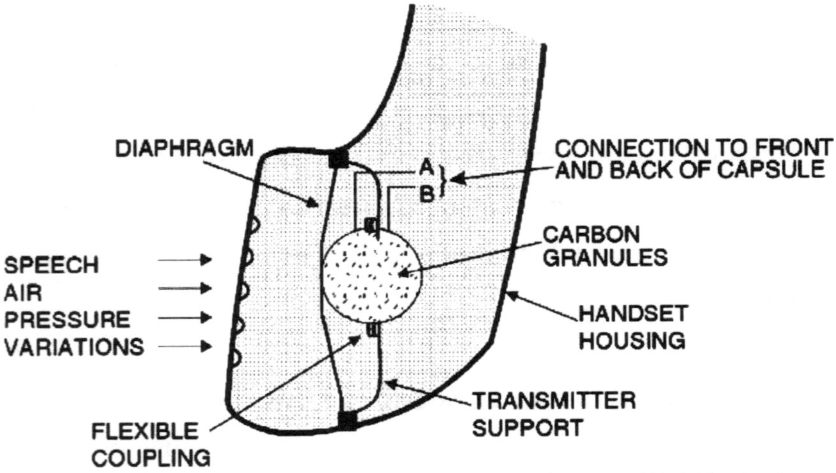

Courtesy abc TeleTraining, Inc. Geneva, IL

diaphragm is released, it momentarily snaps out farther than its original position. The carbon is under less pressure than normal and the resistance of the transmitter is momentarily greater. The current flow decreases.

The diaphragm of a transmitter is made of lightweight phosphor bronze, duraluminum or a similar material. Either the center is strengthened by an extra inner cone of the same material or it is corrugated to act as a stabilizer. The flexible outer edge is securely clamped in the transmitter housing. This design enables the diaphragm to move in and out at the center like a piston. Since the diaphragm is sensitive to sound waves, the carbon granules are compressed and released as the corresponding pressure from the sound wave's changes.

The telephone transmitters in use today are, in principle, like the ones invented more than 100 years ago by Thomas Edison. Most newer electronic telephones use real microphones connected to related speech processing equipment to vary the line current. Small microchips allow economy and space saving, enabling inexpensive, high quality "throwaway" telephones. The output now generated by microchip based telephones must emulate the same variations created by the carbon granule type of transmitter.

What is known as the basic *500 set*, a single-line telephone like the one that was in use in most homes, has dictated the industry's electrical standard for the telephone instrument and all related signal processing equipment.

All types of 2- and 4-wire circuits are still designed around that 500 set.

Note: The 500 set refers to the rotary phone. The same telephone with touchtone is called a 2500 set (see Figure 2.1).

- **The *receiver*** (Figure 2.6). The receiver is the "mouth" of the telephone in that it speaks into the ear of the person using the telephone. It also contains a diaphragm whose movement is caused by the strengthening and weakening of the field created by the magnet within the receiver. The receiver converts the varying electrical current representing the transmitted speech signal to variations in air pressure perceived as sound by the human ear. An electromagnetic receiver consists of coils of many turns of fine wire wound on permanently magnetized soft iron cores that drive an armature. The *armature* is a diaphragm made of a soft iron material.

  When someone speaks a word into a transmitter, the current flow in the circuit is alternately increased and decreased as the moving electrode moves in and out of the carbon chamber. A requirement for an electromagnetic receiver is a

Figure 2.6

The Receiver

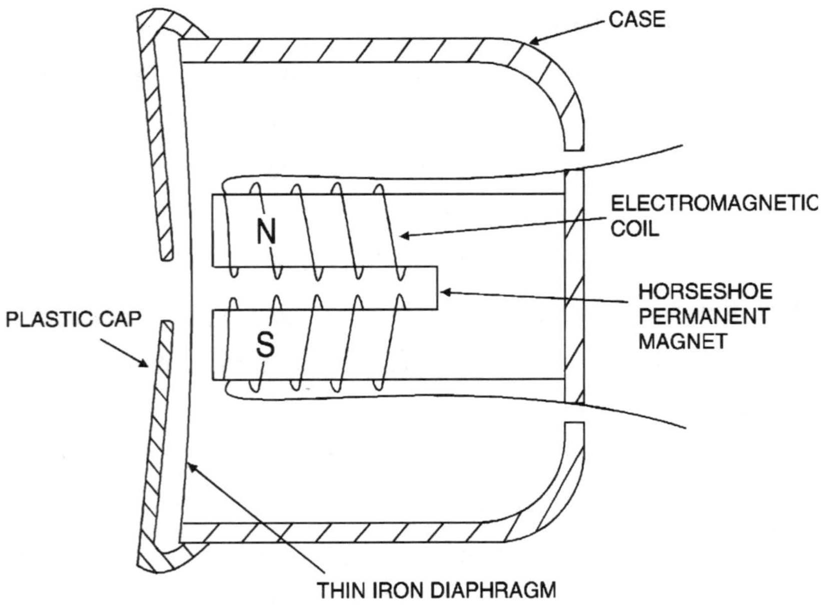

permanent magnet to provide a constant bias field for the varying electromagnetic field to work against. Otherwise, both positive and negative currents would push the armature in the same direction. The varying electrical current representing speech flows through coils and produces a varying electromagnetic field. It alternately aids and opposes the permanent magnetic field; thus, it alternately increases and decreases the total magnetic field acting on the diaphragm. This causes the diaphragm to vibrate in step with the varying current and moves the air to reproduce the original speech that caused the current changes.

Other types of receivers operate similarly, except that the armature is a separate part and is connected to a conical non-magnetic diaphragm. The rocking action of the armature causes the aluminum diaphragm to vibrate to reproduce the original speech. In newer telephones this receiver is created with the use of microprocessors.

The electromagnetic receiver was a central element of *Alexander Graham Bell's* original telephone patent.

Part of the design of the telephone handset that enables you to hear your own voice while talking is called *sidetone* or *side noise*. The reason for this is to give you some feedback that the telephone is working. Too much sidetone causes an echo.

- **The *ringer*.** There is a wide variety in types of ringers. Telephones run on DC (direct current) where electrons flow in one direction. The bell or ringer operates on AC (alternating current), which means that electrons are moving in two different directions to activate the bell. This AC sent on the local loop (telephone line) is called *ring generator* (90 to 105 volts AC at 20-Hz). Minus 48 volts DC is always on the line, which is used to operate the telephone after it is answered.

You may want to review a book such as Lee's *ABC's of the Telephone* (order from 1-800-LIBRARY) for some basic electronic concepts relating to telephone signals, which are electrical signals. There is a good analogy for understanding electrical signals. Envision a garden hose. The hose represents the wire. The water is the current. The water pressure is the voltage (electrical pressure). Stepping on the hose with your foot is equivalent to resistance on an electrical circuit.

- ***Microprocessors.*** The microprocessors in electronic telephones may replace any of the above internal components and may also add additional capabilities and functions to the telephone, such as speed dialing, etc.

Many telephones look the same, but there is wide variation in price. You can buy a throw away single-line telephone for less than ten dollars or a multi-line multi-featured telephone to work with a business telephone system for six hundred dollars. As with many manufactured items, the quality of the components varies and this is reflected in the price. The price also tends to be higher on the proprietary telephones which work with a specific manufacturer's system, even though they may look the same as those you buy in your neighborhood telephone store.

Digital vs. Analog Telephones

In the beginning of this chapter we mentioned that some telephones are digital and others are analog. The distinction has to

Figure 2.7

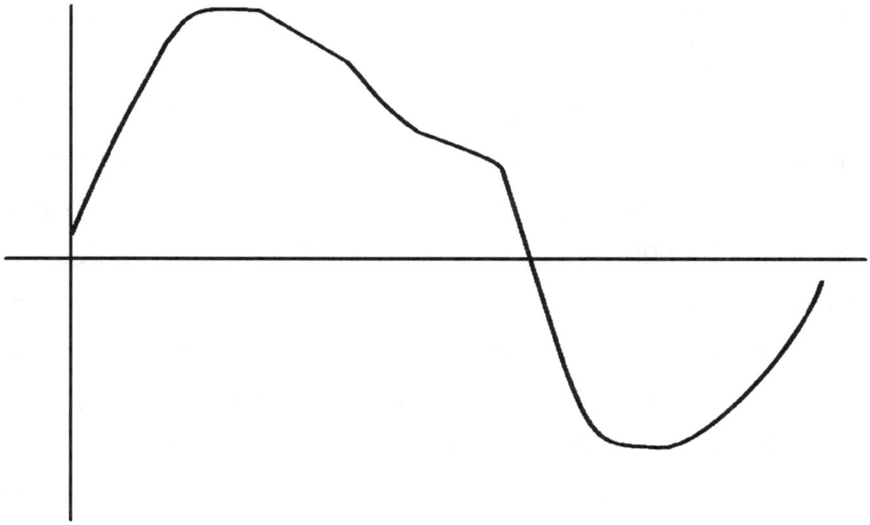

Courtesy of Siemens provided by Clem Napolitano

Figure 2.8

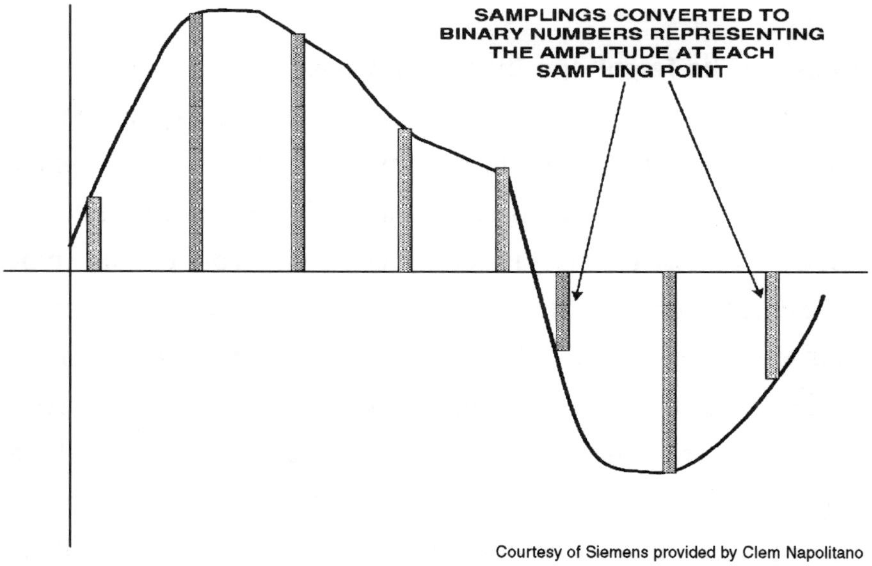

Courtesy of Siemens provided by Clem Napolitano

do with how the speech is processed in the telephone before it is sent over the line back to the business telephone system or central office.

Voice starts as analog, represented by a sine wave (Figure 2.7). In many systems, the voice is transmitted in the analog form. In systems that convert the voice to a digital signal, the voice signal is sampled 8,000 times per second. Each sample is assigned a numerical equivalent, some combination of zeroes and ones. The binary numbering system uses only zeroes and ones to represent all numbers. Each sample is a voltage reading which is given a numeric value and changed into binary form (Figure 2.8).

When the signal reaches its destination, it is converted back to analog so that it can be heard by the human ear.

Telephone systems themselves are also either analog or digital, but this has to do with how the signals are moved around within the system control cabinet. It is possible for a digital telephone system to have analog telephones. Analog telephones may be similar in function and appearance to the digital telephones.

Defining The Telephone By What It Does

Another way of defining the telephone is to explain what it does. An important thing to remember is that what a telephone does depends in a large part on what it is connected to. A telephone is useless unless it is connected to a network of some type.

Some telephones are connected via cable in the walls to a business telephone system (either a key system or a PBX whose control cabinet is located on the business' premises). Some telephones in businesses or residences are connected via cable in the walls to cable in the street and to the local telephone company central office where central office-based switching equipment is located.

Here are a few of the more common functions that many telephones can perform. [Our focus is on business telephones. Residential telephones may have some of the same capabilities which are dependent upon the capabilities of the telephone company central office switching equipment from which the residence receives dial tone.]

Outgoing Calls

Placing a telephone call to an outside location: On some business telephone systems you dial 9, wait for an outside line dial tone, and dial the telephone number. You then hear a ringing signal that is sent to you from the local telephone company central office to let you know that the other end has not yet answered. It is not the actual ringing of the telephone you have called. On other systems, when you pick up the handset you have an outside line dial tone and you need only dial the telephone number, without dialing 9.

This is a good place to define the terms *NPA*, *NNX* and *NXX*. NPA stands for *numbering plan area*, or *area code*. There are over 200 area codes (and counting) in the United States (including Alaska and Hawaii), Canada, Bermuda, the Caribbean and northwestern Mexico. Within an area code, no two telephone lines may have the same seven-digit number. The middle number of the area code used to be a "1" or a "0." This has now changed to create more area codes as the need for telephone numbers increases. This is due to the proliferation of fax machines, computer modems, beepers and cellular telephones. Area codes will now be NXX (N= any number from 2 to 9; X= any number from 0 to 9). This new scheme is known as the *North American Numbering Plan.*

The local telephone company central offices, or exchanges as they are sometimes called (the first three digits of the seven-digit telephone number is the exchange), have been referred to as NNX. This means that only the last digit can be a "1" or a "0." To obtain more numbers, these also are changing to NXX so that the middle digit may also be "1" or "0."

In order to accept these new numbering plans, modifications are needed in many business telephone systems, central office switching systems, and long distance carrier switching equipment.

Intra-office Calls

Placing a call to another telephone inside the office: Most business telephone systems enable you to reach any other telephone on the system by dialing the three- or four-digit extension number of that telephone. The telephone may ring differently to indicate an intra-office call. The display of the ringing telephone may indicate the name of the person who is calling. There are many different types of business intercom communications. Some are separate intercom groups for a specific department, or may be two-way *"boss-secretary" intercoms.*

Some intercoms have *voice announce* which enables the voice of the person who calls you on the intercom to speak to you and you to speak back, without your having to lift the handset. *Off-hook voice announce* (less common) enables someone within your company to speak to you through the speaker of your telephone while you are on an outside call.

Dial intercoms are subgroups within a business telephone system, enabling members to call each other by dialing one or two digits without having to dial the extension number. The call rings on a separate button on the telephone.

Some business telephone system intercoms have *paging*, enabling an announcement to be made from the speaker of every telephone at the same time. Other systems enable access to a separate overhead paging system.

Some smaller business telephone systems have more flexibility in terms of the internal communications options than do the larger systems.

Incoming Calls

Receiving a telephone call from either outside or inside the office: If someone is calling you, you answer by picking up the handset and saying "Hello." Some systems enable you to answer by just pressing a speaker button and saying "Hello" without lifting the handset. Other systems enable you to just say "Hello" without touching anything (seldom used).

Now that incoming telephone calls are providing the telephone number of the caller with *Caller ID*, organizations are providing information about the caller on your computer screen as you answer. (See Chapter 12 on Computer Telephony for more on this.)

On Hold

Putting a call on hold: Many telephones are equipped with a button, often red or orange in color, which enables you to put a call in progress on hold. This means that the call is still at your telephone. The caller cannot hear you, so you are free to do other things such as call someone else, take another call, search for a file or gather your wits. Multi-line telephones almost always have a hold button. Some single-line telephones have one as well.

Some telephones have *hold recall*, which signals you with a tone when you have left someone on hold too long.

Other telephones have *individual hold*, or *I-hold*. This means that if you put a call on hold at your telephone, no one else in the office who has the same line or extension can take the call off hold from any other telephone.

When a call is on hold, the light on the telephone where the call is holding rhythmically flickers to distinguish it from a call in progress (usually a steady light) or a new ringing call (light flashing on and off).

Transferring

Transferring a call: If you are talking to someone on your telephone, you may wish to send the caller to someone else who is on the same telephone system. Most telephones will let you transfer that call so that the caller is now connected to someone else and your telephone is free to handle another call. It is good procedure to announce that you are transferring a call to the person to whom the call is being sent, although the call will still be transferred if you simply hang up after the transfer.

Conferencing

Setting up a conference call among three or more people: Most business telephone systems enable you to set up a conference call from your telephone. There may be a separate button labeled *conference* for this. Usually, the more people you add to the conference, the harder it is for everyone to hear. For conferences of more than four people, it is advisable to use special conferencing equipment or an outside conference service. It is important to know how to drop off one of the conferees from your conference call without ending the call. Not all systems can do this. If you have set up the conference call from your telephone and you hang up, you may disconnect the other call participants.

Last Number Redial

Many telephones store the telephone number you just dialed so that if you have reached a busy signal and wish to try again you need only press the "last number redial" button. On some telephones you do not need to lift the receiver to do this. Some telephones have a similar feature called *save and repeat.* This usually takes up two buttons and enables you to place other calls while the telephone still retains the number you want to retry at a later time.

Speed Dialing

Speed Dialing and Automatic Dialing: This capability enables you to store frequently called telephone numbers in your telephone. Then you need only press one or a few buttons rather than dialing the entire telephone number. If you have spare flexible feature buttons on your telephone, these can often be set up to automatically dial a telephone number by pressing just that one button (typically called *Automatic Dialing* or *Autodial*). Or there may be a button for speed dial that you press, followed by a one or two digit code (on the dial pad) which represents the stored number. On many business telephone systems there is *station speed dial,* which is specific to a particular telephone, and *system speed dial,* which is accessible to authorized telephones throughout the system. If the telephone system requires dialing 9 to dial out, you may have to program the 9 into your speed dialing, although some systems are intelligent enough to add it for you.

Call Forwarding

If you are not going to be at your desk, some telephones enable you to forward calls to another telephone either within your office or at an outside location. Many systems can forward your calls to different destinations depending upon whether your telephone is unanswered or busy, and whether the caller is inside the office or outside. As Voice Mail becomes increasingly common as a means of answering your calls, the need to forward your calls to other telephones is diminishing.

Some telephones are set up with a button that, when depressed, will send your calls directly to Voice Mail if you are not at your desk. This prevents the caller from having to wait for your telephone to ring several times before going to Voice Mail.

Computer Telephony Developer Tips

Department managers would like to see where each of the telephones in the department is forwarded and for how long each has been in the forwarding mode. Reminder notices in the display would help the user to remember to "unforward" his telephone when he returns to his desk.

Call Pick Up Groups

This function enables you to answer another ringing telephone in your office even though the extension number that is ringing does not appear on your telephone. This is usually accomplished by pressing a button on your telephone labeled *call pick-up.* If you have a display you may see the name of the person whose call you have picked up, which will enable you to answer appropriately, "Rose Bodin's office," rather than just saying "Hello." *Group call pick-up* lets you answer any ringing telephone in your preselected group. *Directed call pick up* requires you to know the extension number that is ringing and to dial it after pressing the call pick up button.

Computer Telephony Developer Tips

It would be useful to find out who else is in your call pick up group by scrolling through the display on your telephone or PC screen. To find this out now you must request this information from your telephone installation company or telecommunications system administrator. They bring up the call pick up group information on the administrative terminal. It would also be useful to be able to add and delete people from your group and to have certain extensions in more than one call pick up group. This is not possible with today's telephone systems.

Privacy

In business telephone systems, it is customary for more than one telephone to pick up the same extension number or outside line such as a secretary's telephone being able to answer the boss' extensions. Privacy prevents someone else who has the same line or extension from inadvertently cutting in on your conversation. If you want to let him in, you may do so if you have a *privacy release* button. Not all telephone systems are automatically equipped with the privacy feature.

Older Style Telephones Still in Use

You may hear some of the following terminology referring to the telephone that originated within the Bell System and is still in use today:

— A single-line telephone such as one you may have in your home is sometimes called a *CV* or a *500 set* (rotary), or a *2500 set* (touch-tone). It can be a desk or a wall-mounted model.

– The older electromechanical multi-line telephones, many of which are still in use today, were called *K sets* or *key sets*, referring to the keys or buttons. Each has a red hold button. The KV has six buttons. Then there is a K-10, a K-20 and a K-30, each with that number of buttons. The term *call director* is also used to refer to a large multi-line telephone.

Different geographic areas of the Bell System sometimes used slightly different terms for the same telephone, such as *KK6*, which was Southwestern Bell's term for New York Telephone's KV.

Collectively, this type of electromechanical key equipment was known as *1A2* equipment. The hold button and lights on the phone were controlled by circuit boards in the telephone equipment room.

Naming of telephones today is even less standard than in the past. Each manufacturer makes up names for its own telephone instruments. Sometimes the names mean something. For example, in one manufacturer's line, telephones with a "1" in the number have speakerphones, such as the M2616, and those with a "0" do not, such as the M2606. The model number may also have to do with the number of buttons on the telephone.

Labels

A mundane, but important aspect of the telephone is how it is labeled. Some manufacturers provide printed labels to indicate the extension number or feature of each button on the telephone. Other manufacturers do not and the numbers and features are often written in pencil which looks unprofessional. Telephones often start out neatly labeled, but get messy as changes are made to the features or extensions. Many business telephone systems are incorrectly labeled.

Some telephones enable you to replace all button labels with a single sheet which slips under the clear faceplate on the telephone. Others require that you create a single tiny little label and take off each individual button to replace the label. This can be time consuming when making many changes to a group of telephones.

Color Trivia

The single-line telephones which used to be supplied by the local Bell System company pre-1970 came in a variety of colors: white, black, red, ivory, pink, beige, blue, turquoise and yellow. In the 1980's the color choices for business and residence telephones were typically beige or ash, sometimes gray. Now black is back as the new hot color. One manufacturer is making a system with red telephones.

Expanding the Number of Buttons

Some multi-line business telephones can change from ten to twenty buttons by "popping out" the button strips (called *keystrips*) and inserting a panel with more buttons. Other business telephones can add one or more *modules* to add more buttons.

The following chapters will discuss the array of systems and services with which even the most basic of telephones can interact. In the next chapter, we provide an overview of the telecommunications industry.

Ringing

One thing you can't tell by looking at a telephone is how the ring is going to sound. Yet this is a sound you are going to be listening to many times a day for many years.

Telephone system manufacturers differentiate ringing sounds in different ways. One is the "cadence". For example, a call coming in from outside may be "riiiiing – riiiiing" while an internal call may be "ring ring – ring ring" (short little bursts of ringing in the same time interval that an outside call rings two long rings). Another differentiator is pitch. Some rings have a high pitch changing to a low and others start with a low and change to a high. Some actually have different sounds.

Different ringing for internal and outside calls is becoming less important as information on the display lets you know whether the caller is internal or from the outside world.

Differentiated ringing among a group of telephones all in the same area is often useful. That way if you are not sitting right at your desk, it is still possible to know that it is your telephone that is ringing.

Most telephones do not enable you to have a different ringing sound for each line or extension appearing on a multi-line telephone.

CHAPTER 3
The Telecommunications Industry

We are not going to go back to Alexander Graham Bell, although the history of telephony is a fascinating story. In the beginning, with multiple telephone companies, you may have had three separate telephone instruments on your desk to call people who subscribed to one of the three different networks.

We begin in the more recent dark ages sometime around 1983. AT&T, known as *Ma Bell*, owned the majority of the local telephone companies in what was known as the *Bell System*. The Bell System handled local and long distance telephone calls and provided the installation and maintenance of telephones, telephone systems and cabling. It wasn't perfect, but it was a single source and everyone knew who to call.

In the *Carterfone* decision of 1969, Tom Carter won the right to connect privately-owned telephone equipment to the Bell System network. Another milestone, in the 1970's, was MCI winning a case to compete with AT&T to sell long distance calling services.

In January of 1984, the culmination of an anti-trust suit over which *Judge Harold Greene* presided resulted in the *divestiture* of the Bell System companies by AT&T and the deregulation of the telecommunications industry. 1984 marked the beginning of more change in the telecommunications industry. New players came in to sell systems and services. (What happened to the Bell System in the U.S. is developing in similar ways worldwide.)

Since 1984, many players have come and gone. There is always a host of entrants with new ideas and products to keep things exciting. The most recent excitement is the entry of the computer industry into the telecommunications arena.

The telecommunications industry includes the following types of companies:

Long Distance Carriers

Long distance carriers, also known as long distance companies or *IXCs (inter-exchange carriers)* sell long distance telephone calls. They also rent circuits that permanently connect two or more offices of a large organization. These circuits may carry voice, data or video signals. The circuits are called *leased lines* or *dedicated lines*.

The big three long distance carriers in the United States are AT&T, MCI/Worldcom and Sprint. Worldcom resulted from the merger of Metromedia, LDDS, Wiltel, IDB and ITT. There are hundreds of smaller carriers that cover the entire U.S. or a specific geographic area. Most carriers handle international calls. U.S. companies may have relationships with carriers for local and long distance calls in other countries, since a call may be carried on circuits of several different companies before reaching its final destination. As more mergers and acquisitions take place, having one company handle a call or circuit end to end will become more common.

Even in the U.S., a long distance telephone call may use the circuits of the local telephone companies at both ends of the call. The long distance companies pay the local companies for the use of these circuits.

If you're renting a circuit from a long distance carrier connecting two of your offices, the local telephone companies at both ends may provide the local leg or *local loop* of the circuit, even though you are billed by the long distance carrier.

Sometimes long distance carriers rent circuits from each other. If your long distance company is MCI/Worldcom, you may be using circuits owned by AT&T.

Some calls may still be transmitted by bouncing the signals off of communications satellites. Transponders on the satellite may be shared by different long distance carriers. The trend is away from satellite communications as the use of fiber optic cable becomes more widespread and the resulting capacity to handle call volumes increases.

Long distance carriers generally refer to their *network* as the collective group of all circuits over which calls are sent or permanent connections are made. The network may include a variety of transmission media including copper cable, fiber optic cable (underground and above ground), microwave communications and satellite communications. The network also includes hardware to provide the media with the capability to transmit.

Long distance carrier networks have switches enabling a large number of people to access a common group of outside lines or circuits. These long distance company switches are sometimes referred to as the *Carrier POP*, meaning point of presence. You may reduce your costs by having a circuit that connects your telephone system directly to this *POP*. The circuit may be rented from the local telephone company although you will probably be billed for it by the long distance carrier. The cost of this circuit is called an *access charge*.

Smaller companies and residences reach the long distance carrier networks in most U.S. locations through *equal access*. Each of the major long distance carriers in your area rents space in the switch at the local telephone company central office. If you have signed up with AT&T as your long distance company, when you place a long distance call, the programming in the central office switch will route your call over the AT&T network.

There are hundreds of companies selling long distance calls, representing a variety of long distance carriers. You can buy long distance calls directly from the carrier or from one of these companies which may be a reseller, rebiller, aggregator or variation of the same concept. These companies are providing bulk buying, representing large numbers of customers, and are therefore able to negotiate lower rates for the organizations in the group.

Since a long distance call has become a commodity with all calls being of equal quality, one must select a long distance company based upon reputation, service and clarity and accuracy of billing. The cost of a long distance call is based upon many variables. These include distance, time of day, duration, total number of calls made (higher volume = lower costs) and special promotions in effect at the time you select a long distance carrier. Calls may also be sold at*postalized rates* which means that the cost per minute will be the same regardless of where you are calling. The term postalized rate is becoming somewhat dated as more long distance carriers move to a fixed cost per minute for calls within the U.S.

Long distance carriers provide *800 numbers* enabling callers to reach you without paying for the call. You pay for it instead. Along with the 800 numbers, you can buy some sophistication in the routing of your calls. For example, you can have one 800 number with callers from each state routed automatically to the nearest branch of your company when they dial the national number. Your 800 number can also be programmed to send your calls to your New York office until it closes. After 5PM EST, it will reroute the calls to your California office.

You can receive reporting of the telephone numbers of the people who are calling your 800 number. This is called *ANI* (pronounced "Annie") or *Automatic Number Identification.* More new services are being introduced regularly . As 800 numbers are beginning to run out, 888 has also been introduced as another *toll free* area code.

Long distance carriers also provide: 900 services, 700 services, credit cards, toll fraud prevention programs, and customized bills and management reports on paper, on CD-ROM or on-line Web access and other related services.

The *Telecommunications Act of 1996* has enabled long distance companies to compete for local service. The local telephone companies, meeting certain criteria, can now sell long distance service. Soon they may all become providers of circuits and calls and the distinction between local and long distance service will go away.

Most major long distance carriers also sell local service, internet access and cellular service.

Regional Bell Operating Companies

The Regional Bell Operating Companies, also known as *RBOCs* or *Baby Bells*, were formed in 1984 as a result of the break up of AT&T. AT&T retained its long distance network and the capability to sell business telephone systems, but gave up the ownership of the local telephone companies which then became part of the newly formed RBOCs. Figure 3.1 shows the six RBOCs and the geographic areas they cover. Within each RBOC there are separate companies, including the local telephone companies that are still regulated by the state public service commissions or public utility commissions. For example, up until 1994, NYNEX,

Figure 3.1

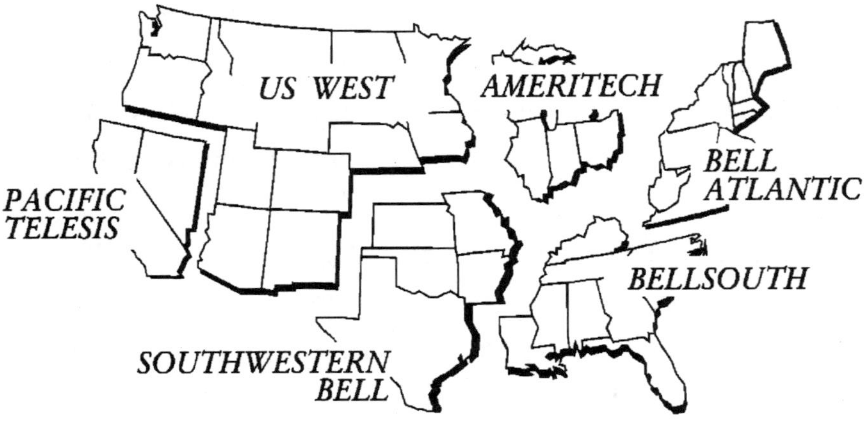

(which was the seventh RBOC and is now part of Bell Atlantic) included the local telephone companies New York Telephone (the NY) and New England Telephone (the NE). The local telephone company parts are still regulated. Each provides local telephone service (primarily dial tone lines), handles local telephone calls and switches long distance telephone calls to the appropriate carrier. Separate companies within the RBOC may sell business telephone systems (not all do). They compete with Lucent Technologies (formerly AT&T) and other telephone installation and maintenance companies. Since their inception, the RBOCs have had to surmount regulatory restrictions to obtain the right to compete in existing markets, including long distance telephone calling and other emerging markets, such as the sale of information.

Most services sold by the local telephone companies continue to be *tariffed.* This entails a review by a public service commission. The tariffs are incorporated into a voluminous set of written service descriptions and prices. Any time there is a price increase or a new service is introduced it is subject to review by the public service commission (some states call this the Public Utilities Commission).

The presence of these regulated companies has hampered the RBOCs in their attempts to be competitive in other telecommunications businesses.

Independent Telephone Companies

There are many telephone companies covering geographic areas that were never a part of the pre-1984 AT&T network. Therefore they are not now part of the RBOCs. These independent telephone companies continue to operate within the boundaries of the area otherwise controlled by the RBOC. GTE is one of the larger ones. The independent telephone companies provide local telephone service (dial tone lines) and may sell or rent telephones to residence and business customers.

The local telephone companies that are part of the RBOCs cannot sell or rent telephones. This is due to the *MFJ (modified final judgment)*, a set of rules that governed the AT&T divestiture and resulting breakup of the Bell System.

The independent telephone companies are regulated by state public service commissions as are the local telephone companies that are part of the RBOCs.

Local Telephone Companies

(also called Local Exchange Carriers, LECS or CLECS)

The local telephone company is what most people still refer to as "the Telephone Company." It is also known as the *Local Exchange Carrier,* LEC for short. There are now also *CLECs*, which stands for *Competitive Local Exchange Carriers.* When you want to order one or more local telephone lines, you call the local business office. You speak to a representative trained to take orders for very specific services. For example, if you call to order a line you may be asked whether you want loop start or ground start, measured rate or flat rate, listed or not listed and on what type of jack the line is to be terminated. Do you want them to run the cable to the jack or will another company do the cabling? If they run the cable, they may charge you a monthly fee to maintain it.

Mountain Bell, C&P Telephone, Illinois Bell — these are a few of the local telephone companies.

The local telephone company sells a variety of services. Many sell the same services under different names. For example, Centrex, Centrum, Centralink, Essex and Intellipath are all essentially the same service, central-office based switching. They are given different names by each local telephone company.

The *central offices* are the locations for large *central office switches* which, combined with the cables and other transmission media providing circuits, comprise the network of the local telephone company. When you order outside telephone lines, in most cases, they are delivered from the closest central office. This is the *serving central office* or *C.O.* Most of the central office switches in the U.S. are manufactured by either AT&T or Northern Telecom. Some are from Siemens, NEC, Ericsson and others. Just as with the PBXs that provide the switching for business telephone systems, these central office switches are of many different vintages and software releases. This affects the type of services that you may order from the local telephone company in your area. For example, you may want to order a service that is technologically possible to obtain from the central office switch. If there is insufficient demand you may not be able to get it, since it may cost the local telephone company too much to provide it on a limited basis.

The tariffs also regulate whether or not you may order a particular service. Even though the central office switch may be capable of providing it, if the service has not yet been approved by the public service commission, you may not be able to get it.

In the late 1960's, local telephone companies began to offer advanced services, sometimes called *Custom Calling Services* or *Totalphone*, to residential customers. These included *Speed Calling* (use two digits to call a frequently dialed number), *Call Forwarding* (send your calls to your neighbor's house while you're over there having coffee), *Conference Calling* (have a three way conversation with Grandma and Aunt Helen) and *Call Waiting* (don't miss important calls – you can take a second call without ending your first call). As it was then and still is today, residential customers often get the more advanced services first although businesses ultimately pay more for the same services. A few more advanced services are now being added which are essentially the same four that were offered in the 1960's. For example, if your phone is ringing and you do not reach it in time, you can press *69 to call back the person who was calling you. These are called *PhoneSmart services* in some areas.

Ordering Local Telephone Service

Typically the local telephone company is set up in the following manner. The *business office* order department is the place to call to request outside telephone lines or to make changes to your existing lines. A business office representative trained to take orders will handle your call and will provide you with pricing, an order number and a due date. We suggest confirming this in writing. You need to do the writing. The telephone company does not send a written confirmation. Obtain the address and telephone number of the person with whom you placed the order. You may be able to fax or e-mail your order confirmation. Call back on the day the order is due to be sure it is scheduled.

It is also a good idea to confirm orders to discontinue service in writing. Always keep a copy of your correspondence with the local telephone company. Keep it for years! If you find at some future point that the billing has continued on this service, having the letter will enable you to request a refund.

There may be two separate business offices, one for residence and one for business. Each office may be further segmented into an order department and a billing department. The business office serves a specific geographic area which is usually defined by your telephone number exchange or NNX, the first three digits of the local telephone number (not the area code). Sometimes the business office is called *COG* which stands for *Centralized Operations Group* or the *BCSC* for *Business Communications Service Center*. This varies depending upon your geographic area.

The telephone number for the correct business office can be found on one of the pages of the telephone bill. If you do not have a telephone bill, call directory assistance and ask for the telephone number of the business office handling the three-digit exchange. For example, "Please give me the telephone number for the Bell Atlantic business office handling the 686 exchange to place an order for business service." The business office handles only local service. It cannot provide information about your telephone system or long distance service, except for the name of the long distance carrier to which your local lines connect.

If you are placing an order for a large number of lines or for some of the more complex services, the business office may take your name and telephone number and tell you that a marketing representative or a sales person will call you the following day. Most local telephone companies have sales departments for handling the larger orders, including many of the business orders. Most very large organizations (more than 500 people) have a designated Account Representative from the local telephone company.

Once you have placed an order, it is transmitted from the business office to various departments. The plant department or installation department will ultimately be responsible for dispatching an installer into the field to complete your order. It's a good idea to obtain the name and telephone number of the field foreman who is responsible. There is also an inside plant department, handling the technical aspects of delivering the service, but not dispatching personnel to the field. Some requests do not require a field visit, such as a request to change your telephone number.

The engineering department is responsible for making sure that the facilities are available to deliver the service you requested. This may include the expansion capacity and programming capabilities on the central office switch, the route from the central office to your premises and the cable from the street into your building and to your floor if it is a multi-story building.

The directory department handles the listing of your telephone numbers in directory assistance that ultimately go into the printed telephone directory. Be sure you don't inadvertently get a new directory listing for your company the next time you order a new telephone number for a fax machine.

There may be a separate company, such as R.H. Donnelley, publishing the directory or just the advertising in the directory. If you have directory advertising and make changes to your telephone numbers, be sure that the changes are reflected in the advertising as well.

Once your telephone service is installed and something goes wrong (even if it was installed several days before), the department to call is Repair Service, sometimes called the *RSB* or *Repair Service Bureau.* A separate group of technicians (not the installation technicians) handles the telephone line repairs. In many telephone companies, you reach this department by dialing 611. Once you have placed a repair request, it is advisable to get another telephone number for them as well. The reason for this is that you may want to call to check on the status of your repair from a location other than the one where the service problem exists. There are different Repair Service Bureaus handling different geographic areas. The local telephone company has many other departments and a defined hierarchy within each department. Everyone has a supervisor, the supervisors have managers and so it goes on up. These hierarchies are now beginning to flatten out.

Alternatives for Local Telephone Service

The *Telecommunications Act of 1996* permits anyone who can show financial, technical and management capability to set up a local telephone company. It still requires that tariffs be filed with the public utilities commission. You must also negotiate either the ability to interconnect with the local exchange carrier or a resale agreement with them.

In most areas, the local telephone company now has competition from these CLECs.

In some areas, you may also buy local telephone service (lines, dial tone and calls) from the long distance carriers and the cable television companies.

Some large organizations are forming their own telephone companies. This enables them to purchase local lines and calls at wholesale rates from the local exchange carrier, substantially reducing the cost of their local telephone service.

LATAs

You may hear the term *LATA*, standing for *Local Access Transport Area*. This is a geographic area more extensive than the local calling area, to which the local telephone company may carry your long distance calls. The LATA was another by-product of the AT&T divestiture. This enabled the local telephone company to retain some revenue by being able to carry long distance calls within the LATA. Once a call leaves the LATA, they are required to hand it off to a long distance carrier, even though it may still be within the territory covered by that local telephone company. There are now some states permitting the long distance carriers to handle intra-LATA calling.

Telephone System Installation and Maintenance Companies and Manufacturers
(also called Telecommunications Vendors or Interconnect Companies)

Before 1969, the local telephone companies were the providers of business telephone systems. After the *Carterfone decision,* it became legal to connect privately-owned telephone equipment to the AT&T network (at that point AT&T still included the local telephone companies). Thus, the private telephone system industry, the *interconnect industry*, was born. The idea of buying your own telephone system rather than renting it from the local telephone company became more widely accepted as the 1970's progressed. The people who developed this industry are to be commended for their entrepreneurial and creative spirit. Many of them came out of Arcata, Norelco, Litton and I.T.T., four pioneers in the interconnect business.

By 1984, when the AT&T divestiture took place, owning your business telephone system was commonplace. Companies who were still renting from the local telephone companies were spending a lot more money than they had to.

After the divestiture, AT&T won the right to keep the telephone systems that were rented to businesses. This was called the *embedded base*. AT&T continued to rent the systems, but also offered the installed systems for sale, and many businesses bought them. The local telephone companies were granted ownership of the cables in the walls which were used to connect telephones rented from AT&T back to the PBX. The PBX itself was rented from AT&T, all on the same premises. For several years, the local telephone companies collected rent on the cables. They also collected a monthly *investment recovery charge* from their customers to make up for the fact that they had had to give up the telephone systems to AT&T.

The confusion over who was responsible for what fueled the growth of the interconnect industry. It was much simpler and less costly to buy your own telephone system, run your own cable in the walls and call the same company to repair anything in the system.

The term *interconnect company* is now rarely used. Almost every business buys its own telephone system. Many of the earlier interconnect companies have been sold to larger companies. The field from which to choose a business telephone system has narrowed considerably and varies from one geographic area to another. This industry is about to undergo another major change with the advent of Computer Telephony and the Internet (see Chapter 12).

The telephone system installation and maintenance companies in your area can be found in the Yellow Pages. Major manufacturers of business telephone systems are Lucent Technologies (formerly AT&T) and Northern Telecom, each with about 25 percent of the market; Siemens (ROLM), Mitel, Ericsson, NEC, Toshiba, Fujitsu, Intertel, Intecom, Comdial, Telrad, Tadiran and a number of smaller companies make up the other fifty percent.

Some of the systems are installed and maintained by the same company that manufactures the system. These installation arms of the manufacturers are often former interconnect companies who distributed the products of the manufacturer and were then acquired. Some companies are authorized distributors and carry the product line of one or more telephone system manufacturers. In either case, the services you can expect from a telephone installation and maintenance company include the following:

- You can purchase a business telephone system from them.
- They will install the system. Installation includes pulling the cable in the walls; connecting the telephones to the cables; and connecting the control cabinet of the telephone system to the other ends of the cables and to the outside lines brought in from the local telephone company. They will also install the switchboard console, used as a central answering point.
- They may also install cable for your computer network. It does not make sense to have two separate companies (one for telephones and one for computers) running cable back to a central point since a significant portion of the cost is for labor (see Chapter 8 on Cable).
- They will program the system to work for the unique setup of your organization. Programming the telephone system

for the way in which it will be used is as important as buying the right system. Most telephone systems have a set of rules determining how they may be programmed. Programming determines such things as which extensions are picked up by which telephones, which extensions are in a call pick-up group for answering other telephones and what happens to a call when it rings on an extension which is unanswered or busy.

— They will train your staff how to use the telephone system and provide an instruction manual.

— They will handle changes to your system such as installing new telephones, rearranging telephones and making programming changes. If you wish, someone on your staff can learn to make some of the system program changes with a *Maintenance Administration Terminal* or *MAT*. The changes made with the MAT are often called *MAC work* which stands for *moves and changes*.

— They will handle all repairs to the telephone system and the cabling.

— They may represent you to the local telephone company. If there is a problem with one of your outside lines, they will report the problem and follow up until it is resolved.

— They may sell other related systems such as Voice Mail, Automated Attendant and Call Accounting which they will install and maintain.

— They may also act as representatives for the local telephone company or a long distance carrier, from whom they receive a commission for selling certain types of services. This is called being an *Authorized Agent*.

— Telephone installation and maintenance companies are not the place to call for problems with telephone bills or for help in deciding the types of local and long distance telephone services you need (unless they sell these services).

Some companies can provide service nationally, but most serve smaller areas. When evaluating a company, it is important to focus on the support available in the local area where the telephone system will be installed.

Some smaller telephone systems (for less than 20 people) can be purchased in telephone stores or through catalogs like *Hello Direct* (phone # 1-800-444-3556). It's a good idea to have an experienced telephone installer put the system in for you rather than doing it yourself, unless you want to learn by doing.

Specialized Telecommunications Equipment Companies

While many telephone installation companies also sell *Voice Mail, Automated Attendant, Call Accounting* and *Facilities Management* systems, another group of companies sells *only* these systems without actually selling the telephone system. The advantage of buying from these companies is that they tend to have more in-depth knowledge of the types of systems they sell than the telephone installation and maintenance companies do. Also, the systems that they sell are often more technologically advanced than the systems sold by the telephone installation and maintenance companies, particularly the larger ones.

The disadvantage is that you now have an additional vendor. If you are buying from one of these companies selling peripheral equipment only, be sure that the systems are already working with the type of telephone system with which you plan to use them. Also, be sure that both companies have worked together or are willing to do so, so that you're not left to do a lot of coordination between them during and after the system installation.

Telephone Installation and Maintenance Companies for Specific Types of Business Operations

Some of these companies have the capability to install and maintain the systems while others sell through telephone installation and maintenance companies.

Trading Turret Companies

There is a group of companies who manufacture a type of telephone system called a *Trading Turret System*, sometimes called a Dealing System. These are large multi-button telephones, usually 60 or 120 buttons, analogous to a key system, but designed with the brokerage trader in mind. Traders require instantaneous communication with a large number of other traders. This is accomplished through point-to-point circuits that appear on the buttons of the trading turret.

The newer versions of these systems are more dependent upon software which provides the appearance of a multi-button telephone on a computer screen. This enables access to many "pages" of different outside lines through which the trader can scroll.

As with telephone system manufacturers, some companies have installation and maintenance arms while others sell their products through distributors who install and maintain the systems.

The companies selling trading turrets include IPC, British Telecom, V-Band, Siemens, Hitachi, Etrali and Positron.

Automatic Call Distribution Systems for Call Centers

Several companies manufacture telephone systems specifically designed to handle large call centers, such as airline reservation operations or groups of order takers. These systems are specialized switches designed to handle high volumes of calls. They route the calls to different groups of customer service representatives and provide management statistics. These statistics include how many calls are handled by each person, how long callers wait to be answered, and how many callers are on hold at any given time.

Many PBXs can be set up to work as an ACD. Separate ACD manufacturers include Rockwell, Aspect and AVT (who purchased Telecom Technologies).

Many Computer Telephony applications are being set up in conjunction with an Automatic Call Distribution System (see Chapter 5 on Automatic Call Distribution Systems).

Cabling Companies

The telephone installation and maintenance company will install the cable for both the telephone system and the computer network.

Other companies are in the business of running cable only. Particularly on larger systems (500 people and up), there may be some financial advantage to having a separate cabling company run the cable. The company from whom you are buying the telephone system may sub-contract the cable pulling to one of these companies anyway.

These are often *electrical contractors.* When using electrical contractors for telecommunications cabling, it is important to be sure that they have experience installing this type of cabling.

If a separate company is installing the cable, be sure that they will certify the work and that the company installing the telephone system (or computer system) on that cable will accept it. The telephone installation company may wish to charge extra for testing the cable, sometimes called *toning and testing*. Certification of cabling typically adds five to ten percent to the price, but it's a good insurance policy. There are different levels of certification, so it's important to be very clear on what is being requested and what happens when something goes wrong.

No matter how small the project, it's a good idea to put cabling specifications in writing and obtain competitive bids. You may wish to take a bid from the telecommunications installation and maintenance company and one from an electrical contractor to see how they compare.

An improperly installed cabling job can create problems with systems that may never be resolved.

Telecommunications Management Software Companies

With multiple telecommunications vendors providing a variety of systems and services and with expenses rising, the need to manage telecommunications within the organization is increasing. There is a small group of companies who write and support software to assist organizations with the management of telecommunications assets and often computer assets as well.

Comware Systems, Inc. (Stamford, CT), Stonehouse and Telco Research are the companies focused on the larger organizations with other companies writing software for smaller systems.

These software systems keep track of the system configurations (circuit boards and spare capacity), the actual desktop devices such as telephones and PCs, the cabling, and the circuits from both the local telephone company and the long distance carriers. In addition, they may track work orders so that as changes are made to the system, all information is automatically updated. Company telephone directories are also generated from the software.

They provide cost allocation and chargeback capabilities for telecommunications equipment, services and calls (see Chapter 7 on Call Accounting and Facilities Management Systems).

Consulting and Telecommunications Management Support Companies

Other companies benefiting from the complexity of the telecommunications industry and the decisions facing businesses are those providing consulting and telecommunications management support.

These companies can be found in the Yellow Pages in your area under *Communications Consultants* or *Telephone Support Services,* or through professional organizations. They do not sell telephone equipment, local or long distance telephone service. A consulting firm will represent your interests and should provide you with independent, objective advice on your telecommunications requirements.

In addition, these companies will help you to manage projects and may provide telecommunications management support such as regularly reviewing the telephone bills and representing your interests to your telecommunications vendors.

Local Area Network Installation and Maintenance Companies *(VARS - Value Added Resellers)*

With the onset of Computer Telephony, you may now buy a telephone system that links to your computer network from the company who sells you the computer network. The number of such companies will increase as more organizations discover ways in which to integrate computers and telephones in order to improve their operations and customer service.

These companies may also be called *Telephone System Integrators* or *Value Added Resellers (VARS).*

As the industry evolves and Computer Telephony becomes more widespread, the participants and types of companies will change.

It is the opinion of one respected industry observer that the future of the telecommunications industry is in the hands of three different types of companies who succeed at the following:

1. Selling information
2. Providing transport of information
3. Providing professional services to pull it all together

The next section of this book talks about business telephone systems and some peripheral systems that work with them.

PART II

Hardware

CHAPTER 4

Telephone Systems: PBX and Key

This chapter explains the functions and operation of a telephone system.

Most organizations use either a key system or a PBX, with the control equipment for the system located on site. With the advent of Computer Telephony, some of these systems are now capable of taking instructions from a PC. The PC may link to a specific telephone within the system or with the entire system at its control point. A PBX can actually reside on a circuit board within a PC. You can read more about this in Chapter 12 on Computer Telephony.

Some organizations use Centrex service with the telephone system functions controlled off site by the central office equipment of the local telephone company.

Regardless of the form that the telephone system takes, there are capabilities that people have come to expect from their systems. Those responsible for developing, setting up and managing telephone systems need to understand these expectations.

As we stated above, most telephone systems are either Key Systems or PBXs. *PBX* stands for Private Branch Exchange. You may also see the terms *PABX*, Private Automatic Branch Exchange, and *EPABX*, Electronic Private Automatic Branch Exchange. Some manufacturers, in order to distinguish their products, use the term CBX, Computerized Branch Exchange, or IBX, Integrated Branch Exchange. There's also NBX for network branch exchange. These are all essentially the same thing.

In terms of function, you may also hear *Switch*, Switching System or the phone system, also referring to a PBX. The "private" in PBX means that the control equipment is on your own premises, as opposed to the "public" switched network.

System Design

The way in which a telephone system is set up is often called the *system design*. Up until the 1980's, the differences between Key Systems and PBXs were clear. *Key systems* were for smaller businesses, and their distinguishing characteristic was that each outside line (outside telephone number) appeared on everyone's telephone.

Key System Design

The lines were (and still are) ordered from the local telephone company. When the first line is in use, incoming calls ring on the second line and on down through the sequence of telephone numbers. This is known as the *hunt group*. It continues until reaching the last line in the group. With all lines in use, the caller hears a busy signal.

If the receptionist answers a call, it is put on hold and the called person is told to *pick up on line* 3. When the button is depressed for the third line at any telephone in the office, the caller is there.

Users of the key system press down a *key* or button of a line to place an outgoing call. They hear the dial tone coming from the local telephone company central office. There is no need to dial 9 to place an outside call.

Many key systems are still set up in this manner and are referred to as a *square* key system (Figure 4.1). This means that every telephone looks the same and picks up the same group of outside lines.

This works well in offices where there are no more than ten outside lines. After that, it becomes difficult to remember who is

Figure 4.1

A Square Key System

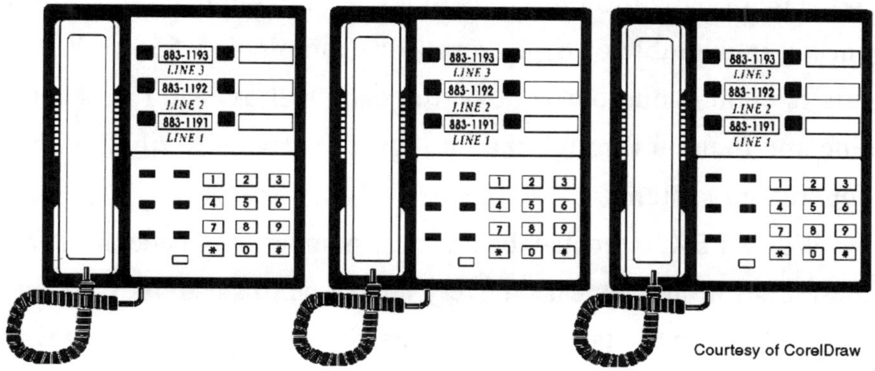

Courtesy of CorelDraw

on which line and is confusing for both users and callers left on hold and asked several times, “Who are you holding for?”

If a company has been operating with a square key system with more than ten lines and it is working well, there may be no need to change if you replace the system.

Key systems usually have an *intercom* so that the receptionist can announce a waiting call to the person requested by the caller. The intercom also enables system users to call each other. Key systems typically have more options in terms of internal communications than do PBXs.

You may hear that a telephone system is “just a key system,” used in a derogatory manner to suggest that it is not as good as a PBX. In fact, most of the key systems currently on the market incorporate many, if not all, of the functions of a PBX, so the distinction has become blurred. You may see the term *hybrid* referring to a key system that can be set up either like a square key system or a PBX. When selecting a telephone system, it is more important to focus on the system capabilities for features and growth than on whether it is a hybrid key system or a PBX.

PBX Design

A PBX has traditionally been set up in the following manner. The organization has a main telephone number, sometimes ending in a double or triple zero (for example 635-5300). When this telephone number is busy, the calls roll over to the second line and then on through the hunt group as we described with a square key system.

With a PBX, there may be a greater number of outside lines and these lines do not appear on every telephone. Instead, they are answered by a central position known as the *switchboard* or *attendant console*. The caller tells the switchboard operator or attendant the name of the person called. "May I please speak to Linda Storaekre?" Then the operator *extends* the call to that person by dialing an *extension* number assigned to the called person's telephone.

To place an outgoing call, the same group of outside lines is used. Since the person placing the call has only an extension number on his telephone, it is necessary to press the button for that extension number and dial an *access code*, usually 9. In doing so, a free outside line is selected and the person can then place a call.

If all lines are in use, as with a square key system, callers to the main telephone number hear a busy signal. People trying to place outgoing calls, when dialing 9, will also hear a busy signal (sent from the PBX, not the local telephone company central office) indicating that all lines are in use.

It is important to understand that the total number of telephone calls, whether they are incoming or outgoing, cannot exceed the total number of outside lines. If you have ten outside lines and receive four telephone calls at the same time as six people are placing outgoing calls, the next caller to your telephone number will hear a busy signal. Outside lines are also called telephone numbers, dial tone lines trunks and *POTS* lines (plain old telephone service).

What we have described above, with all calls coming into a switchboard operator, is a traditional telephone system design (Figure 4.2). With the increasing use of Voice Mail and Automated Attendant, many systems are being set up differently. For example, a type of outside line called a *Direct Inward Dial* trunk may be used. This allows each person using the system to have a separate telephone number to give to callers so that their specific extension may be called directly without the intervention of the switchboard operator. If the directly-dialed extension is not answered or is busy, the call may then go to the switchboard operator or it may be answered by Voice Mail. The caller hears the voice of the person called. "This is Rick Luhmann. I'm not at my desk right now, but please leave a message and I will return your call, or dial 0 for immediate assistance." The caller may then leave a message that will, in turn, activate a message waiting indicator on the telephone of the person called (Figure 4.3).

An Automated Attendant can take the place of the switchboard operator answering the main telephone number with an announcement such as the following: "Thank you for calling Miller Freeman. If you know the extension of the person you are calling, please dial it now. To order a book, dial 1; to place advertising, dial 2; for a company directory, dial 3; or please wait for assistance." You still need a live person at some point to handle callers who do not know what they want or who do not wish to use the automated system (Figure 4.4).

Read more about Voice Mail and Automated Attendant in Chapter 6. They are separate systems, but related to the PBX and often housed inside the PBX cabinet.

Figure 4.2

Traditional Telephone System

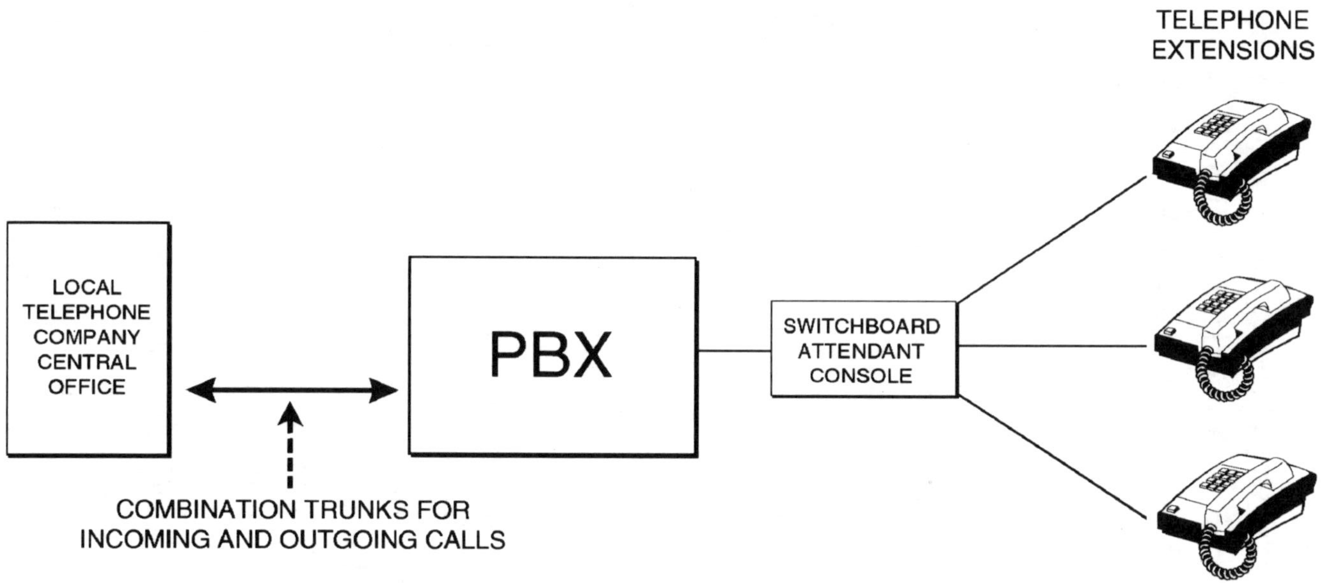

A call comes in on a combination trunk and is answered by the switchboard attendant. Caller is then sent to the correct extension.

Extensions dial 9 to get dial tone from the combination trunk for placing an outgoing call. Switchboard attendant is usually not involved in this process.

Figure 4.3

Telephone System With "DID" Trunks and Voice Mail

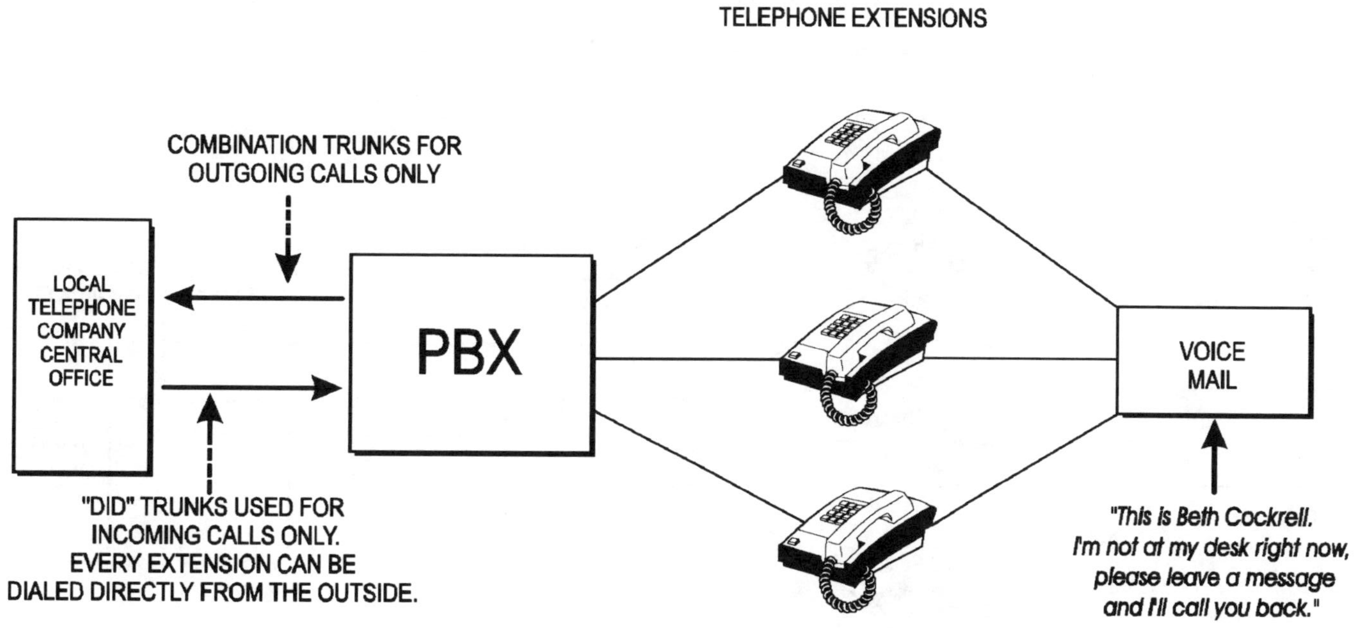

In this scenario calls are placed directly to the telephone extension. When the extension is unanswered or in use, the calls forward to the Voice Mail.

Figure 4.4

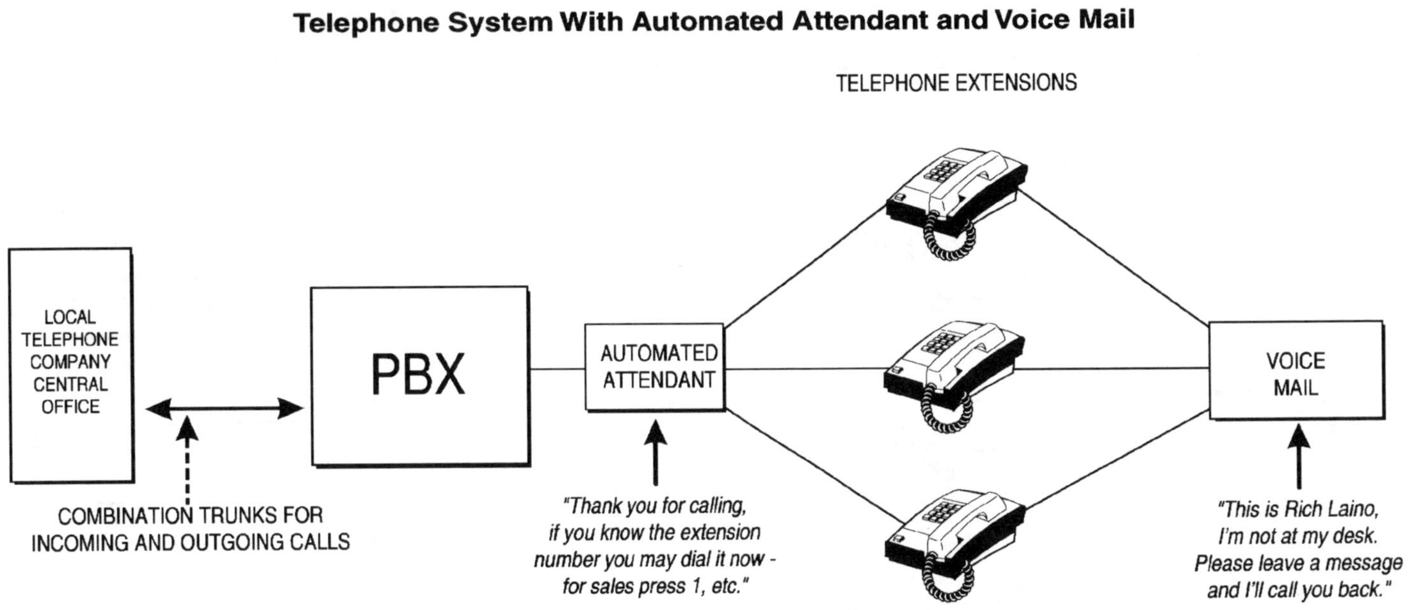

The PBX, Automated Attendant and Voice Mail are housed in the same room and sometimes within the same cabinet. The call movement is controlled by the system software.

Extensions dial 9 to place outgoing calls on combination trunks.

Telephone System Capabilities *(see Chapter 2 for more)*

Now we'll talk about some of the functions and capabilities of the key systems and PBXs.

Answering Position/Attendant Console

Every telephone system has what is known as the answering position where a receptionist or switchboard operator has historically answered the incoming calls. With a key system, the answering position may look the same as each of the other telephones in the system.

Most key systems have a separate module which is attached to the telephone at the answering position known as a *DSS/BLF* or *direct station select* and *busy lamp field.* It consists of buttons representing each of the telephones in the system. When a call comes in, the receptionist answers it. He then looks at the BLF. If the lamp is lit next to the button representing the telephone of the called person, the receptionist knows that person is on a call and asks the caller to wait or takes a message.

If the lamp is not lit on the BLF, the receptionist depresses the button next to the "busy lamp," which automatically rings the telephone of the called person. This is where the name Direct Station Select comes from. It means that by pressing that button, the station (or telephone) represented by that button is "selected."

Some PBXs use the same DSS/BLF concept, but most have what is called the *Attendant Console* or *Switchboard Console* as an answering position. There may be more than one attendant console. Each system has a limit to how many consoles can work with the system. The console is typically much larger and different in appearance than the telephones. Every telephone system's console works differently, but in general, the calls ring in on either one or several buttons on the console (sometimes called *loop keys*), usually accompanied by an audible ring and flashing indicator. The attendant presses the button to answer the call. The caller is then "extended" to the appropriate extension number, which the attendant dials on a touch-tone pad on the console. It is typically necessary to hit another button to send the call to the extension after the extension number is dialed.

Most consoles also have a display which provides some information such as the calling name or extension (for internal calls) or the PBX trunk number that the call is coming in on.

Even with systems that use direct inward dial numbers and voice mail, there is still typically at least one attendant console which also handles calls returning from unanswered telephones, from voice mail and calls from extensions dialing "0".

Some attendant consoles may require up to a 25 pair cable instead of the usual 1 or 2 pair needed for the telephone.

Privacy

Square key systems, where everyone in the company picks up the same group of outside lines, can be likened to a department within a larger company using a PBX. Many people within this department may have the same group of extensions appearing on each of their telephones (Figure 4.5). In this case the system may be equipped with *privacy*. This prevents someone else who picks up a telephone with the same line or extension from cutting in on a conversation in progress.

Some systems have *automatic privacy*, while others require that a separate button on each telephone be activated to ensure privacy. A separate button called *privacy release* can let a person at another telephone in on a conversation if you want him to join you or listen in.

Figure 4.5

Telephones On A PBX With Same Extension Appearances

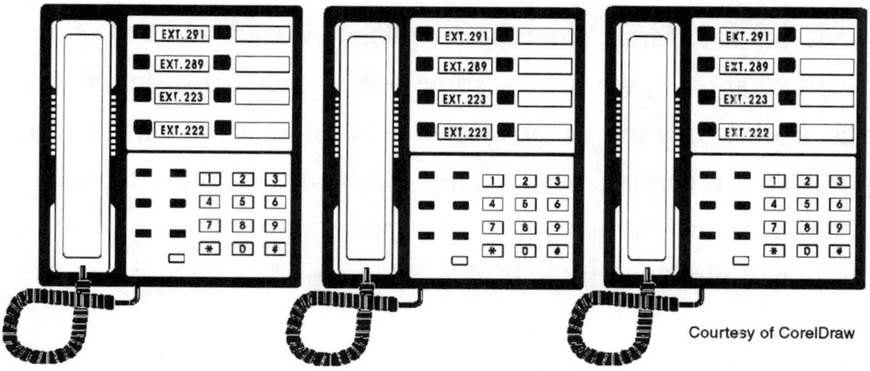

Courtesy of CorelDraw

Call Transfer

If you wish to send a call to another telephone within your office and the extension the call is on does not appear on that other telephone, you must *transfer* the call. Most telephones are equipped with a transfer button that you press prior to dialing the extension number to which you want to transfer the call. You announce that you are going to transfer the call to the person at the other extension, and when you hang up, the call is transferred. If you do not announce the call and the extension to which you have sent the call does not answer, the caller may end up back at your desk, at the switchboard or in Voice Mail. This depends upon how the system is programmed.

Conference Calls

Most telephone systems are equipped with a *conference* button on the telephone. This enables you to set up a conference among three or more people, connecting people within your office to others outside the office. Systems vary in the number of inside and outside callers that can be conferenced. Typically, it becomes hard to hear on a conference call with more than three participants unless you are using specialized conferencing equipment that is separate from the telephone system.

In Chapter 2, we review some other functions of telephones that work with a telephone system. This includes placing an outgoing call, receiving an incoming call, dialing another telephone on the system, last number redial, speed dialing or automatic dialing, call pick-up, and putting a call on hold.

Expense Control Functions

In Chapter 2 we pointed out that a telephone is only a vehicle for using the network. The same can be said of the telephone system. If the key system or PBX is not connected to the outside world, then it is nothing more than an intercom system. Most telephone systems are connected to the outside network. This is also called the *PSTN* which stands for Public Switched Telephone Network to distinguish it from a private switched network, or from the Internet network.

Every time a telephone call is made, money is being spent. Many of the system functions help an organization to control this expense.

TOLL RESTRICTION

Toll restriction enables you to program the telephone system so that each extension has what is called a *class of service.* Each class of service designation enables the extension to call certain areas and restricts it from calling others. The most sophisticated systems can be programmed to restrict the dialing of specific telephone numbers. Others can restrict by area code or by area code and exchange. Figure 4.6 shows a typical scenario.

You may also use class of service to give each extension access to certain system functions and restrict it from using others. You may hear the term *six digit toll restriction*, meaning that the first six digits dialed (the area code and exchange) can be restricted through programming. *Ten digit toll restriction* enables you to restrict dialing to a specific telephone number.

Figure 4.6

Classes of Service For Toll Restriction
(Example)

CLASS 0 =	Internal calls only
CLASS 1 =	Local calls only to 212 area code
CLASS 2 =	Local calls to 212, 718, 914 and 917 area codes
CLASS 3 =	Local and long distance calls within NY metropolitan area 212, 718, 914, 203, 860, 516, 908, 201 and 917
CLASS 4 =	Local and long distance calls to Continental US
CLASS 5 =	Local and long distance calls to US and Canada only
CLASS 6 =	Unrestricted, can call anywhere in US or International

AUTOMATIC ROUTE SELECTION

Most PBXs and some key systems have the capability for *Automatic Route Selection (ARS)* or *Least Cost Routing (LCR)*, which are essentially the same thing. Many larger telephone systems have separate outside lines that, when used, will result in a lower cost for certain telephone calls. For example, your company may have a direct line (dedicated line) connecting you to a long distance carrier. When a call has been dialed from your telephone system, the ARS recognizes where the call is going and sends it over the lowest cost route, based upon how the system has been programmed.

In the 1960's and 1970's many companies had *WATS lines* (Wide Area Telephone Service). WATS lines were separate outside lines enabling calls to be placed at a discounted rate to specific geographic areas within the U.S. (Band 0 through Band 5, as the areas were called.) It was during this period that the ARS was most important. Now the decisions are simpler.

Many companies have just one group of lines for all calls, but ARS continues to be included as a system capability and makes sense to use in certain instances. With local telephone service competition heating up, there may be a resurgence in the need for ARS. Organizations may decide to use different local carriers for handling different types of calls. For example, callers using their computers to dial the Internet through the PBX may be directed via ARS to a group of ISDN lines (see Chapter 9). The advent of making voice calls on the Internet itself may also use ARS.

Telephone systems without ARS can get around this by assigning *access codes* to separate groups of outside lines called *trunk groups*. In this scenario, you may dial 9 for local calls and dial 8 for long distance calls to access a separate group of lines.

CALL ACCOUNTING *(also see Chapter 7)*

Most telephone systems also have the capability to provide information on the calls being made through the system. This includes the time the call was placed, the duration of the call and the telephone number dialed. This is called *SMDR* (Station Message Detail Recording) output, Call Accounting output, Call Detail Recording (*CDR*) output, or *AIOD* (Automatic Identification of

Outward Dialing), an out-of-date term seldom used. It's important to understand that the PBX provides the raw data only. In order to do anything with the data, it must be captured and processed into a usable format.

In many organizations, the SMDR output is stored in a *buffer* device that then is *polled* by a *call accounting system*, usually on site. We will go into more detail on call accounting systems in Chapter 7. Call Accounting is usually PC-based. It accepts the call information from the PBX. It then assigns a cost to each call that approximates the true cost of the call and sorts the calls by extension number. The costs for each extension may also be grouped into department reports, often provided to each department manager.

Calls may also be sorted by *account codes*, numbers which are associated with certain clients or projects of the company. Thus, a law firm will know how much is being spent on each client and may use the information for billing call charges back to the client. Under these circumstances, the PBX requires the dialing of an account code before each outgoing call is made. The PBX keeps the account code associated with the call and sends this information to the *buffer device*.

TRAFFIC STUDY

Most PBXs provide information on how the outside lines in the system are being used. This is called a *traffic study*. Although the call accounting system is capable of providing a traffic study, it is not always set up to do so. Instead, the telephone installation and maintenance company is customarily called upon to poll the PBX for a period of time, often a week, to provide the traffic information. When requesting a traffic study, it is important to be very specific about the information you want and what your objective is.

You may have had a new telephone system installed with 20 outside lines for which you are paying $30 each per month (20 x $30. = $600). Now it's a year later and you want to find out if you really need all of those lines.

A traffic study runs for one week and tracks all incoming and outgoing calls. The results indicate that at the busiest times of the day, no more than 10 of the lines are ever in use. Being conservative, you decide to leave 15 lines in place and remove the other 5 (5 x $30 = $150), thus saving your company $150 per month. *Note*: When you remove outside lines, remind your telephone installation and maintenance company to reduce your maintenance cost, since it is often based upon the number of ports used for outside lines. You must also have your telephone system reprogrammed so that it will not be looking for the missing lines when someone dials 9 or, in the case of your DID trunks, when a call comes in.

Here's another situation for which you may wish to run a traffic study. Callers to your organization complain that they are reaching a busy signal. The one-week traffic study shows that all 20 outside lines are in use ten percent of the time, confirming that during this time callers cannot reach your company. You order 5 more local outside lines from the local telephone company and have them connected to your system by the telephone installation company. You may need to buy an additional circuit board for the system to accommodate these new lines. Several weeks later you run another traffic study that shows that at the busiest times of day, no more than 22 lines are in use. Callers cease to complain about busy signals.

It's a good idea to have a traffic study run on your telephone system at least once a year. You may negotiate an annual traffic study as a part of your PBX maintenance contract. Or you may learn how to run the traffic study in-house with your call accounting system.

MAC (MOVES AND CHANGES)

Many companies spend thousands of dollars moving telephones around the office and changing the extensions or features on the telephone. This activity is MAC work (MAC = moves and changes). It can be provided by your telephone installation company. Most PBXs enable you to do some of your own MAC work using a PC with software that makes program changes to the PBX. In some older systems, the changes may be done with a keyboard only or a dumb terminal.

For example, two executives are exchanging offices. It is necessary to change the extensions appearing on the telephones in each of the offices, on the secretaries' telephones, and on the telephones of the groups backing up each of the secretaries. If the telephones are all in place, these changes may all be done from the on-site MAC terminal, without calling in the telephone installation company. It is always advisable to inspect the individual telephones to be sure that the changes took effect and to change the paper labels which most telephones still use. (*Note*: Some newer systems enable users to take the telephone with them and plug it in in the new office, retaining all of the extensions and features.)

Another example of a PBX change which can be made onsite is that of changing the toll restriction on a telephone. Your call accounting system may indicate that a lot of long distance calls are being made from the lunchroom telephone. Since employees are permitted to make local calls only, that telephone is now reprogrammed so that only calls to the local exchange can be made. All other call attempts will be routed to the switchboard attendant.

An advantage of having a MAC terminal on site (and someone in your company trained to use it) is that it will provide you with current information about how your system is programmed. Otherwise, you may be dependent on your telephone installation company or may even need to collect the information manually. MAC terminals may enable you to obtain the following:

— A list of all extensions in use in your telephone system.

— A list of all spare (unused) extensions in your telephone system.

— A list of all outside lines (trunks) in your telephone system. This will be by the trunk number assigned to each outside line in the PBX. It will not provide the actual telephone number assigned to the line by the local telephone company.

— A company telephone directory listing name, department and extension number.

— A list of all extensions with their associated class of service, meaning which areas can be called and which system functions can be accessed.

— A list of all classes of service in use in the system and what they mean.

— A list of how each extension is programmed to forward calls under a variety of conditions, including when the extension is busy and when it is not answered.

— Information on how each of the slots and ports (see next section of this chapter) is used in the PBX control cabinet.

— Identification of spare slots and ports available for expansion.

Understanding Your Own System

In terms of thinking about what a PBX can do, it is important to remember that each manufacturer's PBX is different and has its strengths and weaknesses. One PBX (or key system) may be outstanding in some areas but not in others. When working with a particular PBX, the more you understand it, the better you will be able to program and use it. Each telephone system installation company has people on staff who understand the PBX their company sells. If a company sells or supports too many different PBXs, the likelihood of staff knowing any one of them well diminishes.

Most telephone installation companies' knowledge is based upon the collective experience of their staff. No one person knows everything there is to know about a particular PBX. There is just too much to know and it changes regularly, as products are continually upgraded and refined. The manufacturers must do this in order to remain competitive. What the telephone installation company does not know, you may be able to find out from the PBX manufacturer.

If you are responsible for managing a telephone system, you may wish to join a *user group*. These groups exist for all of the major PBXs. People using systems in real world environments are often the best source of information on a particular system.

Hardware Inside The PBX Cabinet

Next, we're going talk about the PBX itself and what it looks like. When referring to the *telephone system*, we include the telephones, the cable and the PBX control cabinet, but here the focus is on the cabinet itself. The equipment housed in this cabinet connects the telephones within the office to the telephone lines and the world outside the office.

Figure 4.7 shows a typical PBX cabinet. Key systems and hybrid systems are assembled in a similar manner.

PBX Cabinet

The PBX cabinet is a metal housing designed to hold the electronic components that make the PBX work. In some cases, you buy the entire cabinet which can house equipment for the system at its maximum capacity. In others, you may buy modules which stack up or sit side by side. In smaller systems, the cabinet may mount on the wall.

When planning for the installation of a PBX, it is important to find out the exact dimensions of the PBX cabinet (and dimensions of any expansion cabinets that may be added) to allow for sufficient space. Additional space is required around the cabinet for people working on it, for ventilation and for doors which may swing open (some are removable). To ensure that the requirements for electricity and grounding are met, you need to know where the PBX plugs in. The outlet must be located at the correct height.

Other things to know are the BTU per hour output (PBXs should be kept in a similar environment to computers) and to be sure that the floor can support it, the weight. *Note:* Some newer PBX's are controlled by one or more cabinets that are essentially PCs with telephony circuit boards. (See Chapter 12 for more on this.)

PBX Shelf

Each cabinet or cabinet module contains one or more shelves. Lucent Technologies (formerly AT&T) calls the shelf a *carrier*, but other companies call it a shelf.

Figure 4.7

Appearance of Typical PBX Cabinet

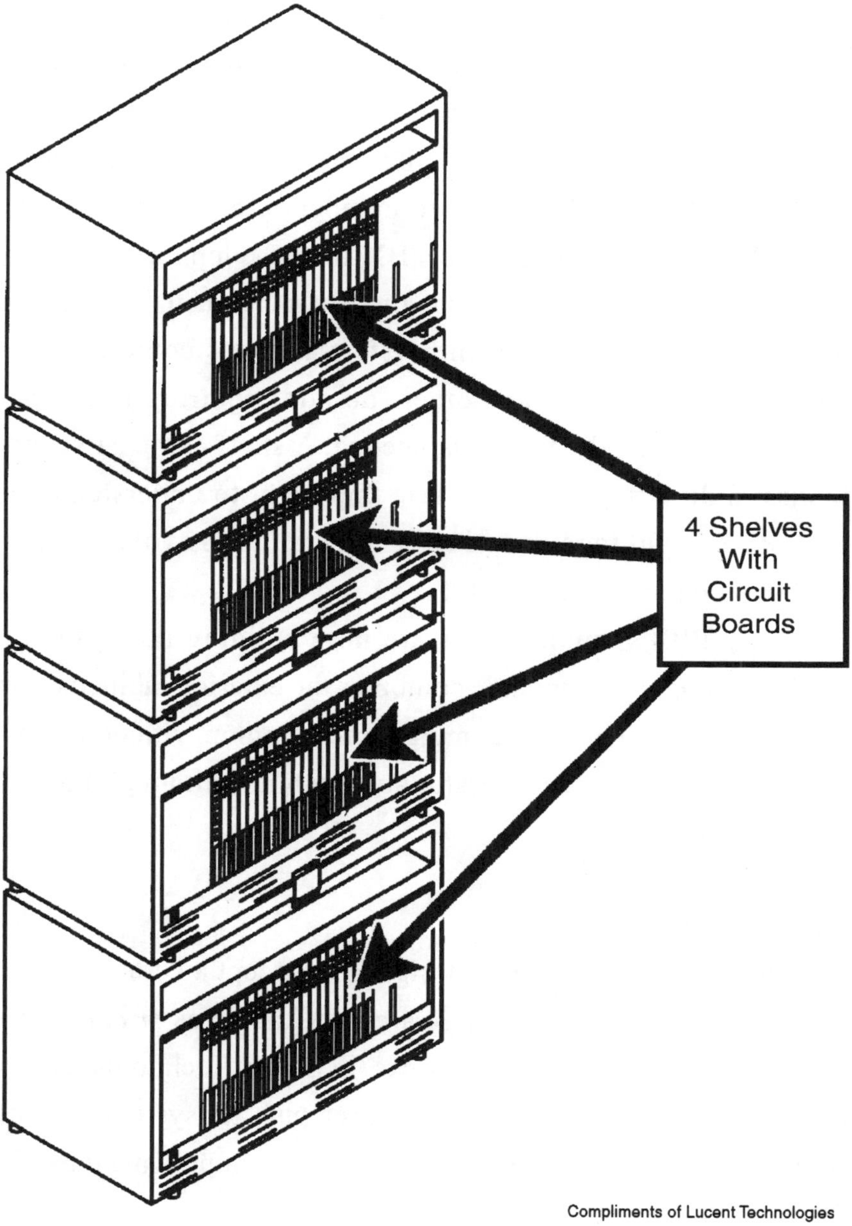

Compliments of Lucent Technologies

The shelf is made up of a group of slots into which fit the circuit boards. Most systems have different *slots* for different types of circuit boards, although some systems may have *universal slots* which can accept more than one type of circuit board.

Each PBX has a numbering scheme so that the location of each shelf and slot (and cabinet, if more than one) can be identified. The numbering scheme also includes the actual ports on each circuit board. *Note:* These numbers are typically labeled on each cable connecting the PBX to the telephones. For example 01-04-08 may represent the 8th port on the 4th circuit board on the first shelf. Some system cabinets come equipped with all of the shelves and others do not, requiring future purchase of additional shelves as needed. A shelf can cost several thousand dollars or more, so it is important to understand how each PBX is put together when it is installed to avoid surprises when you choose to expand it.

Many PBX manufacturers are now building the system in stackable *modules*. A module combines the exterior cabinetry with the shelf inside. This is becoming more common with newer systems than the older "refrigerator size" cabinet, although both are still in wide use.

PBX Circuit Boards

Each shelf in the PBX is designed to accept a certain number of printed circuit boards. These are also known as *circuit cards*, or just *boards*, *cards* or *circuit packs* (in Lucent Technologies systems). Just as each PBX manufacturer puts its system together somewhat differently, the names of system components may also differ, so it is important to clarify terminology.

There are different types of circuit boards within the PBX:

BOTH-WAY TRUNK CIRCUIT BOARDS

Both-way trunks or combination trunks are outside lines that may be used to receive incoming calls or to place outgoing calls. There are *ports (electronic places)* on each circuit board, one port per trunk. Some systems have sixteen trunks per circuit board. Others have eight, four or two. A few older systems use two ports per trunk instead of one.

Every time you add outside lines, you need to consider whether or not you have sufficient ports on the circuit board to handle them. If you do not you need to buy another board. Trunk circuit boards can run from several hundred dollars up to several thousand.

Sometimes both-way trunks are used for outgoing calls only and may be referred to as *DOD (Direct Outward Dialing)* trunks.

DID TRUNK CIRCUIT BOARDS

DID (Direct Inward Dial) trunks are a special type of trunk used for incoming calls only. They enable the direct dialing of each individual extension in the PBX. You may have ten DID trunks for 100 separate DID telephone numbers. DID trunk circuit boards handle either sixteen or eight trunks in most systems. Note: DID trunks delivered on a PRI ISDN circuit may also be used for outgoing. This has created the oxymoron *two-way DID.*

UNIVERSAL TRUNK CIRCUIT BOARDS

A universal trunk circuit board enables you to mix both-way trunks and DID trunks on the same board. This creates efficiencies in the use of the ports and therefore can lower the cost in terms of the total number of circuit boards to be purchased. It can also save space (sometimes called *real estate*) within the PBX cabinet.

ISDN CIRCUIT BOARDS

Some PBXs can accept ISDN (Integrated Services Digital Network) lines. (See Chapter 9 for an explanation of ISDN.) The circuit board may be for a PRI (Primary Rate Interface) ISDN line which has 23 B channels plus one D channel. A few PBXs can also accept BRI (Basic Rate Interface) ISDN lines with 2 B channels plus one D channel. The ISDN lines can be used for voice conversations and data transmissions. They are frequently used to connect a computer to the Internet for higher capacity on the circuit, enabling more data to be sent in a shorter period of time.

TIE-LINE CIRCUIT BOARDS

Tie-lines are point-to-point lines connecting two PBXs so that the users of both systems may communicate without dialing an outside call. Some systems have separate circuit boards for these tie-lines. Most systems use a circuit board which handles two or four tie-lines.

T-1 CIRCUIT BOARDS

A T-1 is a high-capacity circuit using two pairs of wires, enabling the transmission of up to 24 voice conversations at one time. In order for it to do this, a piece of hardware called a multiplexer is required at each end.

The T-1 circuit board is a multiplexer that fits right into the PBX shelf. In some older PBXs the T-1 circuit card must be placed on a shelf specifically designed to handle it. If you are going to use a T-1, you may need to buy a separate shelf for it. Each PBX has limits. Find out how many separate T-1s your PBX can hold. You may hear the term *channel bank* which is a multiplexer external to the PBX which can be used if your PBX does not have T-1 circuit board capability.

DIGITAL TELEPHONE CIRCUIT BOARDS

Most telephone systems now use digital telephones. This means that the analog voice signal converts to a digital form right in the telephone and goes back to the PBX cabinet as digits (combinations of ones and zeroes). For every 8,16 or 32 of these telephones in the system it is necessary to have a digital telephone circuit board. This may also be called a *digital station board* or *digital line board.* Each port on the board corresponds to a specific digital telephone in the system. Every telephone has an associated numeric location in the PBX cabinet indicating the port, the circuit card and the shelf.

ANALOG TELEPHONE CIRCUIT BOARDS

Most telephone systems are installed with at least one analog circuit board with ports for either 8 or 16 analog telephones. Many single-line telephones are analog. The voice signal is sent from the telephone to the PBX circuit board in an analog form.

The analog ports provide analog extensions required for using fax machines and computer modems through the PBX.

There are differences of opinion as to whether or not faxes and modems are best run through the PBX or with separate outside lines. Some PBXs may affect the signals in a manner that distorts or slows down the data transmissions.

Most PBX manufacturers will not guarantee throughput of data beyond a certain speed. If you expect to have computers and fax machines go through the PBX, it is important to find out what speeds can be expected. You may decide to bypass the PBX.

Some systems also use analog ports for interfacing with a Voice Mail or Automated Attendant system although smoother integration is accomplished with a digital interface.

DTMF CIRCUIT BOARDS

DTMF (standing for Dual Tone Multi Frequency or Touch-tone) signaling requires a separate circuit board within the PBX with DTMF receivers. If you have to wait too long for PBX dial tone after you lift your telephone handset, you may need more DTMF receivers. The receiver is engaged when someone within the system is dialing a telephone number and is freed up after the dialing is completed.

COMMON CONTROL CIRCUIT BOARDS

These circuit boards house the central processing capability of the PBX. Most PBXs now have common control circuit boards on every shelf, which is a distributed type of processing. Many of the pre-1980's PBXs had the common control on a single shelf only. If it failed, the entire system failed.

OTHER COMPONENTS INSIDE THE PBX CABINET

You may see a power supply, a fan and a variety of other electronic components within the PBX cabinet. The Voice Mail and Automated Attendant may also reside within the cabinet.

It is important to remember that the circuit boards, like the telephones, are specific to a particular manufacturer's PBX. They are not interchangeable with other PBX's boards. They may not even be interchangeable with circuit boards of other PBXs in the same manufacturer's product line.

This does not hold true with some newer PBXs that use circuit boards for switching and voice processing functions that are purchased from a third party and may be used in competitors' systems as well. But even when this is the case, the way each system operates is unique.

Some newer PBXs do not have proprietary telephones, suggesting that any telephone can be used with the system. In this case, it becomes important to look at all the system functions and how easily they operate, since giving up proprietary telephones designed for a specific system often involves trade-offs in simplicity of operation.

How The PBX Works

Now we know what is inside the PBX cabinet that we can see. Next, let's talk a little bit about what is going on in there that we cannot see, that which is controlled on the many printed circuits by the system software.

We mentioned in the beginning of this chapter that the PBX is also called a switch, which is actually what it does. Switching refers to the methods of connecting the calling telephone with the called telephone. It is necessary in all cases where the telephone needs to be connected with more than one other telephone. Most people using telephones want to be connected with many other telephones, but only one at a time. A switching system is the means for establishing many combinations for connecting any two telephones which have access to it. If it were not for switching, you would need a direct permanent physical connection between your telephone and every other telephone with which you wanted to be connected.

Up through the 1960's, many PBXs still in place had a *cordboard* as the attendant console. If you know where to look (old hotels and apartment buildings, for example,) you still see a few today. The cordboard served the basic functions of: enabling the extension user to signal the switchboard operator, enabling the operator to set up a circuit and connecting the extension user to an outside line.

From this cordboard comes the term *tip and ring*. This refers to two separate wires on the plug which, when plugged in, meet with two contacts in the outlet to complete the circuit. This term is still sometimes used to refer to the wires needed to complete the telecommunications circuit. It may also be used as a term for a *POTS* (plain old telephone service) line, which is a regular dial tone line.

The cordboard was the point at which the control of the switching took place, connecting telephones to outside lines in varying combinations by plugging in cords.

As the number of outside lines and telephones grew, along with the need for less reliance on the switchboard operator, newer types of switching systems evolved. These automated the basic operations of the cordboard. This took place in both business locations and telephone company central offices (where cordboards had also been used).

Several generations of electromechanical switching methods followed. The invention of the transistor permitted the use of electronics within the switch. This device, along with the integrated circuit, provided higher speed and lower cost, using less power.

Telephone systems, in both businesses and in the telephone company central office, now resemble computers. The system logic is no longer electromechanical, as in the past. It is now controlled by stored instructions in the system memory. In some older systems (1960's,

1970's and early 1980's) there is a physical path set up through the PBX for every call. This is called a *space division* system. In other systems, the spaces were time-shared by small encoded bits of conversation in what is called *time division.*

Most systems installed today are digital switching systems. As we mentioned earlier, the voice is converted to digital form and sent through the PBX in that form. It may continue out over the network of the local and long distance company in digital form if the circuits are digital.

The memory, along with the *common control*, is the heart of the PBX. The program controlling the basic switching functions of the PBX is held in permanent memory and protected against accidental destruction. The processor uses memory to store and manipulate data when processing a call or performing some other system function.

Diagnostic and maintenance programs are also held in memory to provide automatic troubleshooting.

In a system using centralized logic or control, the central processor performs all of the logical functions. It scans the extensions to determine whether the extension is trying to use the system. It connects the extension to the outside lines. It receives and stores the dialed digits which determine the destination of a call. It completes the connection and ends the connection to the outside line when the extension hangs up. It operates autonomously with little reliance on the other subsystems in the PBX.

Other systems use distributed logic. In this scenario, the central processor manages the subsystems, issuing and accepting commands to provide the same types of functions described above.

Although all PBXs use one or more memory systems, the devices used and the manner in which they interact differ from one system to another. The basic building block of the PBX is the printed circuit board.

PBX Growth and System Sizing

When planning the purchase of a PBX, growth capability is a major consideration. Although they may be modular in design, in terms of telephones and outside lines that can be connected to the system, most PBXs have a maximum capacity. In most systems, one port represents the ability to connect one telephone or one outside line. A 400 port system can accommodate a total combination of 400 outside lines and telephones.

There are no precise ratios in terms of how many outside lines are required for the number of telephones. A conservative amount would be 25 outside lines (for incoming and outgoing calls) for every 100 telephones. If there are not a lot of calls, 10 lines may be sufficient. If it's very busy, more than 25 may be needed.

There are statistical tables and software programs available to enable you to judge the total number of outside lines needed. Using these statistics to determine the number of outside lines required is called *traffic engineering*. In order for this to make sense, you need to know how many calls of what duration will be handled during the busiest hour of the day. Most people do not have this information, and therefore use judgment combined with trial and error in estimating the number of outside lines needed.

In the world of telephone traffic engineering you may hear the term *CCS* which stands for 100 call seconds (C being the Roman numeral for 100) or, more simply put, 100 seconds worth of telephone calls. 36 CCS or 3600 call seconds are equal to one hour worth of calls known as an *Erlang*. The statistical tables or software programs express call volume in this manner to determine the number of outside lines needed to handle this volume of calling. Something called *grade of service* is expressed in terms such as P.01 meaning that, statistically speaking, one percent of all calls will be blocked.

📖 For information on Traffic Engineering see The Complete Traffic Engineering Handbook by Jerry Harder (order from 1-800-LIBRARY or www.telecombooks.com).

Getting back to the number of ports in the system, another thing to remember is that Voice Mail, call accounting systems, paging systems and many other functions also take up ports in the PBX.

The ability of a PBX to grow is dependent on the system software as well as the hardware. You may open the PBX cabinet and see many empty slots, but if the memory in the system software has been used up, you may not be able to add anything else.

Most people do not pay much attention to how the software is used in the PBX until a crisis occurs. It would be better to plan for the use of the system software, just as the rest of the system is carefully planned. For example, knowing what kinds of functions use memory and in what quantity. Many PBXs are set up with significant portions of the memory allocated for system functions that may never be used. All of the system capabilities use memory in some way.

Disaster Recovery Planning

It is also important to plan what to do when things go wrong. *Disaster Recovery Planning* is the term often used for this activity. What if the power fails? Many PBXs operate with separate batteries known as *battery back-up* or *UPS (uninterruptible power supply)*, which will keep the system running from two to eight hours, depending upon how much back-up has been purchased. Other systems have what is known as *power failure transfer.* This enables outside lines to be switched to predesignated single line analog telephones in the event of a power-driven PBX failure.

Find out what to do when the power returns to ensure that your PBX will be up and running quickly.

What if some or all of the processing power of the PBX fails? Many systems are installed with *redundancy* of varying degrees up to and including having a complete duplicate system standing by.

Other types of backup may include two separate sets of outside lines from two separate telephone company central offices (very costly) or two different long distance carriers.

The most important aspect of disaster recovery planning is to document it and let everyone know how the plan is to be carried out. Testing is also imperative.

Trends In PBX Evolution

The following thoughts on trends in PBX evolution were provided by Allan M. Sulkin, a well-known telecommunications consultant who writes and teaches for Business Communications Review magazine.

Many of the capabilities of the telephones now sold with PBXs can be provided on the PC. As the PBX manufacturers are realizing what the marketplace wants, the PBXs themselves are evolving into general purpose communications controllers, capable of transmitting voice, data and video. They are becoming more adaptable and less proprietary.

Some of the enhancements now being seen are:

1. RISC (Reduced Instruction Set Computing) processing methods;

2. Increasing support of high capacity (wideband) communications through the switching matrix;
3. Wireless telephones, integrated with the PBX, working as extensions;
4. FDDI (Fiber distributed data interface) connectivity for LAN communications support; and Ethernet connectivity with TCP/IP (Transmission control protocol/Internet protocol);
5. Video teleconferencing at the desktop via the PC;
6. Support for ATM (Asynchronous Transfer Mode);
7. Support for ISDN.

Most PBXs are based upon a 32-bit main switch processor element. The manufacturers are now increasing their processing capability in anticipation of future applications.

Lucent Technologies' Definity G3r was the first PBX to use an RISC processor as the main element of its PBX. Some of the other manufacturers will follow suit and others will move in a different direction.

In terms of evaluating how the PBX will perform, the processing design is as important as the processing power. There is a trend among some manufacturers toward a distributed processing design. This includes having more intelligence at the circuit board level or at the telephone level, rather than having it reside in the cabinet or shelf controllers.

Having a powerful main processor reduces the need for distributed processors at the PBX cabinet or shelf level. This results in fewer places for things which will affect a large number of telephones or outside lines to go wrong. Reduction in the number of levels in this processing hierarchy places more emphasis on the circuit boards controlling the telephones, outside lines and interaction with Voice Mail.

Most circuit boards are equipped with 8/16 bit processors, which reduces the processing burden on the central processing unit. Thus, when a microprocessor on one circuit board fails, a limited number of outside lines or other terminal devices are affected. The increased intelligence on the circuit boards also improves the capability of the maintenance company to pinpoint the source of problems through remote diagnosis. It enables more customization and flexibility in designing the individual telephones in terms of their access to system functions.

Another aspect of the processing design that improves the system performance in a PBX is sometimes called *Memory Shadowing*. The same concept may go by different names depending upon the manufacturer. As mentioned, some PBXs have redundant common controls set up so that if the processor in one fails, the system will automatically transfer control to the duplicated processor. This is so transparent that calls in process are not even interrupted when the transfer takes place. Many PBXs have redundant control, but it is this *transparent* transfer of the control that makes the memory shadowing a desirable system function.

Another trend in PBX evolution taking place is that of replacing adjunct applications processors with those that are fully integrated with the PBX. The control of formerly separate types of applications such as Voice Mail, ACD (Automatic Call Distribution), wireless telephones and multi-media messaging are now being incorporated into the PBX processor design.

While processing and design are important in terms of evaluating a PBX's performance, as the merging of telephone and computer technology continues, the software operating system and programming language used may be of more significance. They affect how flexible a PBX will be in incorporating future enhancements.

It no longer makes sense for PBX manufacturers to maintain their proprietary main *CPU (Central Processing Unit)* elements and operating systems of the past. Instead, they are opening up their systems to support third-party or customer-specific software development. PBX companies supply *APIs (applications programming interfaces)* for use with adjunct computer systems. The actual PBX software is not typically open, however, so the customers are currently limited to developing applications that run on adjunct processors. You many also hear the term *OAI, Open Applications Interface*, coined by Intecom but adopted by others, to indicate that the PBX has some degree of openness.

It is anticipated that the first breakthrough in customer-generated PBX programming will be at the systems management level. Many customers with both PBXs and Local Area Networks need a single system to administer and control both of these. This need in the marketplace is driving the development of an open, standards-based management and applications platform.

There is a protocol called *SNMP*, Simple Network Management Protocol, which may serve as a basis for this. *Object-Oriented Programming* is another processing tool that can make it simpler for PBX users to interface their operations. It can help system designers and customers develop more advanced applications. *OOPS* may have a major impact on the ability to manage and administer systems, including both system and network reconfigurations. It may also improve system security.

With Object-Oriented Programming, programming commands are replaced with screen based icons and symbols which speed up the interface between the system and the person administering it.

Another PBX trend is to support the switching of increasing volumes of data at higher speeds, ultimately moving toward accommodating *ATM (Asynchronous Transfer Mode)*. In effect, the PBXs will become voice/data/video switches, rather than voice only.

ATM (Asynchronous Transfer Mode) subsystems will support high-capacity channels which may use *SONET (Synchronous Optical Network)* transport and be used for applications such as high-resolution full motion video conferencing, high speed graphics workstations, medical imaging, Local Area Networking support, and Wide Area Networking support. Many of the larger PBX manufacturers such as Lucent, Northern Telecom, Fujitsu, NEC, Siemens and Ericsson are developing ATM switching capability for the public network.

As is typical in the telecommunications industry, different manufacturers have different ideas of how things will work in the future. Some believe that the PBX will continue to provide basic telephony features and low-speed data switching and interface to an ATM switching environment. Others, proponents of the Computer Telephony integration concept, believe that voice servers will replace the PBXs and that voice will be just one of the types of information delivered to the workstation on a Local Area Network.

Wireless

Wireless telephones working with a PBX are based upon adjunct (separate but connected) controllers that use analog links to interface to the main PBX. In the future, this will probably not be necessary as distributed base stations that transmit and receive wireless signals will be supported by separate circuit boards which will reside on a shelf in the PBX cabinet. Allocation of the frequency spectrum and standards must still be resolved before the use of wireless PBX extensions becomes widespread in the United States. This is one area where European and Asian companies may move more quickly since the U.S. regulatory issues do not exist there.

It is important to remember that there is still a substantial installed base of analog telephone systems. Even the digital PBXs installed throughout the 1980's and early 1990's are not capable of supporting many advanced capabilities and computer telephony integration.

The point is that the implementation of much of this new technology will depend on the willingness of business and other organizations to pay for upgrading their systems. Their decision will depend on how useful the applications are perceived to be.

Technology development labs are currently working on *optical switching*, which may make today's circuit switching technology obsolete.

A Telephone System Request For Proposal

Whether you are upgrading an existing system or purchasing a new system, **put it in writing**. This gives you a better chance of getting what you need and avoiding misunderstandings. It also provides a basis for comparing systems.

To conclude this chapter, here is an excerpt from a written description of a telephone system. There are also some questions from a Request for Proposal used for one of our clients at DIgby 4 Group, Inc. We believe that technical specifications should be accompanied by clear English (or whatever your language is) descriptions of how you, the client, want things to work.

Description of the Telephone System Operation

Incoming calls from outside callers will ring into the system in one of several ways.

— A caller dials the main telephone number 212-883-0000, which is answered by the switchboard attendant and extended to the requested person. There will be two switchboard consoles, but sometimes only one will be in operation.

— A caller dials a direct inward dial number which rings directly to the telephone of the person dialed. Each telephone, except for telephones in common areas, will have a direct inward dial number.

— A caller dials 212-983-0000, which is the first of four separate outside lines (not direct inward dial) which appear on the telephones in a customer service department.

Note: The following describes what is known as *Call Coverage*. This means, "How are we going to be sure the phones are covered?" If a person is not at the desk when the telephone rings, the call can be covered in one of several ways.

— Answered by another person who has a direct appearance of that extension on their telephone. Live coverage is a key objective, so coverage of this type will be widespread.

— Answered by another person who has depressed his call pick-up button to answer the ringing telephone, although he has no direct appearance of that ringing extension. The display on the telephone of the person answering the call will indicate the name of the person for whom the call was originally intended, so that the call can be answered appropriately.

— Answered by another person to whom the call has been forwarded by the person who is not at the desk.

— Answered by the switchboard attendant. Our objective is to have most calls handled at the department level rather than returning to the switchboard console.

Intra-office calls are answered by Voice Mail that will provide the caller with the ability to leave a recorded message.

Anyone answering a call, including the switchboard attendant, will be able to offer the caller the option of leaving a message in the Voice Mailbox of the person being called. The caller will then hear the *personal greeting* of the person being called and will leave a message.

All telephones will have the capability to send unanswered calls to Voice Mail. This will be done by each user before leaving the desk, by depressing a button on the telephone. This button may also work to enable redirecting of incoming calls when the extension is busy.

The capability to automatically forward calls from unanswered and busy telephones into Voice Mail must be a part of the system, although we may elect not to use the system in this manner at the outset.

If a person is busy on another call, an incoming call can be covered in several ways:

— The call can "roll over" to another extension which appears on the same telephone and rings simultaneously on one or more other telephones. Please let us know if there is a limit to the number of multiple appearances of an extension.

— Once the call has rolled over, it can be answered by another person by depressing the call pick up button, although there may be no appearance of that extension on his telephone.

— Answered by another person to whom the call has been forwarded by the person at the busy telephone.

— Answered by the switchboard attendant, although, as stated above, it is our intent to have a minimal amount of callers return to the switchboard.

Intra-office calls to a busy extension can be answered by Voice Mail, giving the caller the option of leaving an automated message.

Any caller reaching Voice Mail will have the option of reaching a live attendant by holding on at any point in the process. Instructions for escaping from Voice Mail by dialing 0 are included as a part of the system.

Outgoing calls will be made by dialing 9 to access a group of combination trunks or a T-1 for long distance calling. *Automatic route selection* in the system will select the lowest cost circuit for placing the call.

Internal calls will be made on the system intercom by: (1) dialing a three or four digit extension, (2) on a separate group intercom or (3) on a boss-secretary type of intercom. All three types must be available and work when the person called on the intercom is on an outside call.

Most telephones in the system will be multi-line with a display. The display will have the ability to indicate the name of the person who is calling on internal calls; and the source of the call (trunk identification by type and telephone number assignment – 7 digit) for incoming outside calls. It will also display the name and number of the outside calling party, when available.

Required System Capabilities

The following capabilities are required for the proposed system to be considered. Please state in your response whether each capability is standard, optional or not available in the proposed system. (*Note*: This list was prioritized for our particular client.)

1. Capability to obtain a traffic study, on demand, indicating use on the incoming, outgoing and tie-lines. Also must be able to track recalls to the switchboard from unanswered and busy extensions and internal calls.
2. Capability to make system changes on-site such as relocating telephones (to cabled, system activated locations) or changing line appearances or system forwarding.
3. Capability for incoming calls to forward in four different directions for internal/busy, internal/no answer, external/busy and external/no answer.
4. Capability for T-1 circuit terminations.
5. Capability to accept an ISDN PRI type of circuit.
6. Capability to integrate with an Interactive Voice Response system.
7. Capability to integrate with a Voice Mail system.
8. Station and system speed dialing capability (state number per station).
9. Built-in speakerphone capability on multi-line telephones.
10. Conferencing of up to five people. This must include the capability to drop calls to a busy or unanswered telephone from a conference call already in progress.
11. Differentiated ringing for internal and outside calls and for different telephones within the same area.
12. Toll restriction and the capability to override it.
13. Last number redial.

14. Capability to provide both boss-secretary and dial intercoms which can signal another telephone while it is in use with an audible tone distinguishable from the regular ringing telephone.

Desirable System Capabilities

Please state in your response whether each capability is standard, optional or not available in the proposed system.

1. Capability to generate a printed directory of extension users.
2. Capability to provide some Automatic Call Distribution (ACD)-like reports as an indicator of staff productivity, such as the number of calls handled by each extension and the duration of the calls.
3. Capability to recognize Caller ID and ANI information sent to the PBX from the central office.
4. Capability to retain Caller ID or ANI digits with a call that is transferred or a call coming through the Automated Attendant.
5. Capability for off-hook voice announce.
6. Capability to set up a busy lamp field for the switchboard attendant console and for individual multi-line telephones.
7. Toll fraud security features on the PBX and Voice Mail systems.
8. Capability to use a PC-based telephone as a replacement for a multi-line feature telephone on the proposed system.

Please provide the maximum for the following:

- Number of times the same extension can appear on other telephones.

- Number of extensions in a call pick up group.
- Number of call pick up groups in the system.
- Number of times a call can forward when the original extension is unanswered or busy.
- Number of forwards for which the original destination of the call or the Caller ID will continue to appear on the display of the ringing telephone.
- Number of buttons on each telephone instrument proposed which can be used for extension appearances.
- Number of speed dial numbers per telephone and system wide.
- Number of participant capacity on a conference call.
- Number of ports for each type of circuit board proposed.
- Number of spare slots in the cabinet as proposed.
- Number of ports for telephones and outside lines in the proposed system at its maximum capacity.
- Number of seconds for the set-up of an outgoing call.

System Specifications

Please base your pricing on the following requirements:

- One digital PBX
- One T-1 circuit board for 24 DID (direct inward dial) trunks
- One T-1 circuit board for access to a long distance carrier
- Growth capacity in cabinet for a third T-1
- 40 Combination trunks (growth capacity in cabinet to 60)
- Two attendant consoles with DSS/Busy lamp field.
- 250 Multi-line digital display telephones with a minimum of ten buttons which can be programmed for line appearances

and system feature access (Note: *Line Appearance* refers to a button on which a PBX extension is located)

- 50 Multi-line digital display telephones with a minimum of twenty buttons which can be programmed for line appearances and system feature access
- 8 Digital single line telephones
- Growth capacity in cabinet to add 40 multi-line telephones
- One system administration terminal and printer
- A call back modem to prevent unauthorized access into the remote maintenance port
- One Voice Mail system integrated with the proposed PBX. – 8 ports (Growth capacity in cabinet to 16 ports)
- One Voice Mail system administration terminal and printer (Please state if the same terminal and printer can be used for PBX system administration)

For the above-described system, provide a total price and the components, showing how the price was computed.

Once we have agreed upon the extent of recabling or reuse of existing cabling, please provide a separate price for all cabling related work, separating materials from labor.

Provide Optional Pricing for the Following

- 4 and 8 hour battery back-up.
- Back-up to maintain DID trunks on the T-1 in the event of a power failure.
- A PC-based Call Accounting system to work with the proposed PBX (including hardware and software). Please mention any toll fraud detection capability of your PBX or Call Accounting system.

Additional Information Requested

1. *Add-on* pricing for all system components including stations, circuit boards and additional cabinets. *Note:* The term Add-on is used to refer to any addition of equipment to an original purchase agreement.

2. Maximum capacity of PBX – in existing cabinet, and with additional cabinets.

3. Maximum capacity of Voice Mail – in existing cabinet, and with additional cabinets. Do ports need to be dedicated to either Automated Attendant or Voice Mail, or can the same ports be used for both?

4. Number of hours of memory in the Voice Mail system as proposed.

5. Is a two-way speaker-phone on the multi-line display telephones standard or optional? If optional – what is the cost?

6. Does the proposed system support manual ringdown as well as automatic ringdown private line circuits (both 2 and 4 wire) on the multi-line telephones and the telephone expansion modules?

7. How many pairs of wires does the proposed multi-line telephone require?

8. With your proposal, please include copies of your standard purchase and maintenance contracts.

9. Please enclose a picture of the proposed telephone stations and attendant console.

10. Please provide a description of the intercom options with the proposed system and of exactly how they are operated by the extension user.

11. Is the proposed Voice Mail system capable of incorporating remote fax retrieval capability? If so, please explain how this is accomplished.
12. Does your installation price include all coordination required with the local and long distance telephone companies? Please describe your procedures for providing this support.
13. Does your installation price include training? Please provide detail of the scope of training provided.
14. Does your maintenance agreement include support of the Call Accounting and telecommunications management software, including a hotline for customer questions?
15. Does your pricing consider any "trade in" on the existing PBX?
16. Does your company sell and support Video Teleconferencing equipment? If so, please describe these capabilities.
17. Are the handsets in the proposed system hearing aid-compatible? Can a *TDD* (Telephone Device for the Deaf) be used to call to and from the proposed system?
18. Please provide a list of at least ten customers using the proposed system (same release for PBX and Voice Mail and same system administration software for PBX).

Maintenance Support Requirements

1. Two-hour emergency response and 24-hour standard response is a requirement.
2. The proposed system must have remote diagnostic capability. Please explain your procedure for monitoring the system performance remotely. Can the remote point be deactivated by the customer to prevent unauthorized access?

3. Please provide the point of dispatch for your repair technicians.
4. Does your company reduce the cost of the maintenance agreement if the customer agrees to provide coordination of repairs with the local and long distance carriers?

The above is considered to be a brief Request for Proposal document. Some are much more detailed. The purpose of including it is to give you an idea of some of the questions to ask and the level of detail required to purchase a telephone system.

The next chapter talks about ACD (Automatic Call Distribution), a particular type of PBX.

📖 For more information on Key Systems & PBXs, see Which Phone System Should I Buy? by TELECONNECT Magazine (order from 1-800-LIBRARY or www.telecombooks.com).

CHAPTER 5

Automatic Call Distribution Systems (ACD)

An ACD or Automatic Call Distributor is a specialized type of PBX. It is typically used in what is known as an *Incoming Call Center*, although outgoing calls are made there also.

When you call to make an airline reservation or to order something by mail, your call is likely being answered by an ACD. When your call is answered, you may hear the now familiar recorded announcement, "All representatives are busy with other customers, your call will be answered by the next available representative."

Three characteristics that distinguish an ACD from an regular PBX are:

— Calls are answered by a representative in the order that they are received.

— Statistics are available on the level of customer service and the productivity of the people answering the calls.

— Most business telephone systems are designed to have more telephones than outside lines, the assumption being that not everybody is going to be on the telephone at the same time. For example, you may have 100 people sharing 25 outside lines.

In an ACD environment, it may be just the opposite. A smaller group of people handle a greater number of lines, assuming that callers will wait for some period of time for the "next available representative." With an ACD, 25 people may be handling 100 outside lines. This affects the system components and software, distinguishing an ACD from a regular PBX.

As with PBXs, every ACD works in a different way. No two are exactly alike.

Some companies manufacture telephone systems to work specifically as an ACD and not as a regular office telephone system. Their systems are known as *standalone ACDs*.

Most other PBX manufacturers have ACD capabilities that be purchased as an option. A PBX can be set up to work as an ACD only. You may also have an ACD as a sub-system of an office PBX, but not have all telephones working from the ACD part of the system.

The Collins Division of Rockwell pioneered the ACD in the 1970's with an installation at Continental Airlines and later at Pan Am and United Airlines. The system evenly distributed large volumes of incoming calls for reservations and ticketing. *Note:* This bit of history was contributed by Paul Lutz of Rockwell.

The Call Center environment has been one of the first to take advantage of Computer Telephony. In many instances, when your call is answered by a representative, a computer screen of information about you pops up. This is called a *screen pop*. More about this in Chapter 12 .

ACD Hardware

The ACD is put together in the same manner as the PBX. There is a control cabinet which holds printed circuit boards. There are separate circuit boards for different types of outside lines and other circuit boards controlling the telephone instruments. The system can expand in the same manner as a conventional PBX.

Some manufacturers use slightly different telephones for the ACD *agents* (also called *Custom Service Reps* or *TSRs Telephone Service Reps)*. These telephones usually have a display which may provide information such as how many calls are waiting to be answered. An ACD telephone may also be equipped with a headset, since the calls tend to come in one after another. This makes it cumbersome to keep picking up the handset to say "Hello."

The buttons of the telephone itself look different with an ACD. It is typically set up so that only one call at a time can be answered by each representative, so only one extension number will appear. There may be a separate button used for outgoing calls. *Note*: An ACD is probably not the right choice if your environment requires each individual to juggle multiple calls on hold.

Other buttons on the telephone may enable the representative to *log on* or *off* the ACD system. When the agent is logged off, no calls will be sent to that telephone.

As in other areas of telecommunications, there is a lot of terminology unique to ACDs. The *queue*, for example, refers to the group of callers waiting (lined up in order, controlled by software) to be answered.

Abandon rate is another favorite, measuring the rate at which callers tire of waiting in queue and hang up.

As mentioned above, telephone service representatives in a Call Center are often called *TSRs* or *agents*.

ACD System Features

Here are some of the system features you may find on an ACD. These were provided compliments of Fran Blackburn at Intecom (telephone 1-800-344-1414 or www.intecom.com), a manufacturer of PBXs often used for the large (100+ representatives) Incoming Call Center ACD.

Inbound Call Routing

The heart of any inbound Call Center operation is *Inbound Call Routing*. There are two phases of inbound call routing: routing to a group, and then routing to a specific agent within a group. Inbound Call Routing uses intelligent programming to recognize and accommodate Call Center traffic and agent performance. The system may have the flexibility to:

— Anticipate changing traffic and performance patterns, respond to unanticipated changes in traffic or agent performance and play announcements at any point.

— Route callers to voice processing devices such as IVRs (see Chapter 10), Fax-on-Demand (see Chapter 11) or Voice Mail equipment, always providing the caller the option to return to his place in queue.

— Recognize priority callers.

— Advise callers of the number of callers ahead of them or anticipated time before they will be answered.

— Put a pause in inbound routing until some predefined event occurs or some period of time passes. Queue calls to multiple groups simultaneously, including groups at remote centers and agents at home.

— Permit supervisor, administrator or management to make changes as needed.

— Most Call Centers try to distribute calls evenly among all agents within a group. However, there are instances where some agents are better prepared than others – for example. when new agents are added to a more experienced group. Management wants the better qualified agents to handle more calls than the new agents. Several different schemes can be used for selecting an agent within a group:

1. Top down/bottom up
2. Longest idle
3. Agent priority (based on experience or skill set)
4. A combination of longest idle and agent priority
5. Performance Parameters & Thresholds

To be successful, Call Center management must set performance objectives, and then measure performance against these parameters. Real time displays and historical reports should show actual performance as compared to these criteria. A list of performance objectives might include average:

- Time to answer
- Talk time
- Hold time
- Work time
- Idle time
- Time in queue

Overflow

Several of the performance objectives listed above can be used for more than just retrospective analysis of Call Center function. For example, they can be used to determine inbound call routing.

Overflow is a feature that recognizes when agents in one group are backlogged, then reroutes calls to another group that may have fewer calls waiting. In the best implementations, the traffic/performance analysis is done automatically and continually in real time, freeing supervisors and management personnel from minute-by-minute monitoring of each groups call volume.

Scheduling

Large Call Centers frequently extend their business hours beyond the traditional work day. Staff may rotate through different shifts or part-time employees may be used to provide extended coverage.

At the same time, call volumes tend to be cyclical. Some times of day or days of the week will always be more active than others. Management must provide a way to schedule employee shifts to reflect these changing requirements.

Balancing the number of employees and anticipated call volumes is a time-consuming task. Call Centers frequently purchase *automated workforce management software* to complement their Automatic Call Distribution systems. Another option for maximizing agent productivity is to mix inbound and outbound calling in one center.

Predictive Dialing

Many Call Centers are designed to both take incoming calls and place outgoing calls. The same group of agents may be responsible for both sales and collections. The sales calls are primarily incoming, and the collections calls are placed during lulls in incoming traffic.

Outbound calls may be placed automatically using *predictive dialers*. Predictive dialers analyze incoming call traffic and agent activity, then automatically place outgoing calls when agents are about to be available. These predictive dialers help management increase agent productivity by decreasing idle time. When you receive a call during dinner and there is a slight delay before a telemarketer comes on the line, a predictive dialer was used to placed the call. When the system detects that you have answered, it sends the call to the telemarketer.

Some Call Centers may focus primarily or totally on these outgoing calls, known as *telemarketing*. In these cases, less sophisticated dialers can be used. Regardless of the Call Center orientation (incoming, outgoing or a combination) management require similar features to assure optimal functionality.

Sign-On/Off

Agents may serve more than one group and may rotate among several workstations during a shift or during a week. In order to monitor and evaluate agent performance, it is important to be able to track which agent is at which telephone. Each time agents make a change, they are asked to sign on and sign off the group they are serving. Automatic sign on/off processes save time and prevent abuse of the system. Using Windows-based software programs, supervisors may also move agents from one group to another .

ID/Password

Agents may sign on using an identification number, with or without a password. The ID and password should be viable from any agent workstation, freeing agents to work at any desk during any shift. When agents serve multiple groups, tracking requirements will determine whether or not the agent uses the same or different ID and password for each group served.

Auto Answer/Auto Release

Automatic Answer and Automatic Release are features frequently used together and in conjunction with headset operations.

With Auto Answer, agents receive calls without lifting a handset or depressing a button on the telephone. Agents may be alerted to the incoming call with a ring, beep, message or other indicator. Auto release terminates the call when the local or long distance carrier disconnects. Thus, the agent is immediately freed to move on to other activities such as follow-up paperwork or receiving the next call. This feature also prevents agents from adding seconds or minutes to each call by delaying release. Auto release serves as a productivity tool for the agent and as an anti-abuse measure for management.

Call Alert

Work conditions in a busy Call Center would be unbearable if phones rang with every incoming call. Silent room conditions (where phones do not have an audible ring) improve morale and productivity. Alternative means of alerting agents to incoming calls include:

- Audible ring, ring-beep, *zip tone* (heard only by agent) for agents with a headset.
- A brief announcement, which may relate to call origin or may prompt the agent for a unique greeting. Also used for agents with a headset.
- Visible indicator such as a flashing lamp or phone display, typically with incoming line information.

Screen Pops

Screen Pops can shave seconds from many if not all calls. Using a computer-to-PBX/ACD interface, Screen Pop brings the caller profile to the agent at the same time a call arrives. This saves the agent from having to ask for the caller's name, account number or other relevant information, keying in a data request and waiting for the screen to appear.

Some systems use *ANI* (automatic number identification of the calling number) to initiate the database inquiry and screen transfer, while others use an IVR (Interactive Voice Response System) to ask the caller to input identifying information that will in turn be used for database access and a screen transfer.

Wrap/Work

Call Center managers must be careful to balance pressure for agent productivity against increasing agent stress and frustration. *Wrap* and *Work* are features that allow the agents some period of time between calls to complete call-related paperwork.

Management may predefine the wrap period, or may allow agents to invoke Wrap as needed.

Work is a similar feature, allowing agents time to complete call-related tasks. Unlike Wrap, Work is not an automatic feature. Agents place themselves in and out of Work as needed. While some systems may combine these two features, they should be defined and reported/displayed separately for better evaluation of agent performance.

Transaction Code

In addition to any database information agents may enter into the caller's profile, there may be a need to collect other call-related information. For example, agents may collect information on "method of payment," "how caller heard about the company or offering", etc. This information may be used for specific reports to evaluate marketing activities, accounting requirements, etc.

Agents are able to enter transaction codes at any time during or after the call. Codes should be flexible, allowing for multiple fields, and should provide a display so that the agent can verify the input before completing the data.

Emergency Record

Some Call Centers are susceptible to harassing or threatening calls. It's important to be able to document these calls without alerting the caller. Agents should be able to immediately conference-in a centralized recorder as soon as they recognize a call of this type.

Supervisor Alert

When agents receive threatening, harassing or other problem calls (difficult questions, hostile caller) they may need supervisory assistance. With Supervisor Alert, agents are able to get their supervisor's attention without interrupting the call in progress. Supervisor Alert may have these features:

- Transparent to caller.
- Alerts designated supervisor; then, if not available, hunts for first available supervisor in management defined sequence.
- Informs supervisor of agent calling and reason for the alert.

Calls in Queue Display

Agents need to balance providing polite and friendly service against the need to answer as many calls (or place as many outgoing calls) as possible. A "calls in queue display" notifies agents if there are any calls waiting, and alerts them if the queue builds beyond a management-specified threshold. Thus, agents are prompted to shorten calls during busier periods.

Wall Mount Display

A wall-mounted display of Call Center volume and agent performance can also serve as a motivator for agents. Mounted in positions of high visibility, displays can be used to scroll information about incoming/outgoing call volumes, agents on line, average time to answer, etc. These displays serve as a reinforcer to agents, and provide valuable information for supervisors when they are away from their desks.

Agent Statistics Display

Typically, when agents are able to monitor their own performance they become more productive. Agent station sets can provide immediate feedback for agents who want to evaluate their own productivity. Some of the statistics that might be displayed include:

- Time on line (since signing in)
- Number of Call Center calls handled (inbound/outbound)
- Average time to answer
- Average talk time
- Time in wrap/work idle time (time between calls)
- Number of non-Call Center calls

Supervisor Agent Features

Supervisors may have the option of serving as agents. Supervisor stations may be able to support all agent features in addition to supervisory features from one telephone.

Real Time Displays

Supervisors are responsible for ensuring high agent productivity without sacrificing quality service. The supervisor's most important tool for accomplishing this task is the set of real-time displays of agent and group performance provided. To be most effective, these displays may be color coded and presented in either tabular or graphic formats as a user selected option. Displays may show real-time status as well as historical information.

Groupings

Any one supervisor may be responsible for about 10-15 agents, even though the entire group serving a particular function may be much larger. Supervisors may be able to identify agent information for their specific agents without scrolling through the entire group display. Similarly, supervisors may be able to display a subset of agents from several different groups, if necessary (for example, in evaluating trainees throughout the Call Center). This super group/ sub group capability is very important in larger Call Centers.

Reports

In addition to real time information used for daily supervisory functions, historical reports are necessary. These identify trends and are vital to planning.

Reports may include information about individual agent and group performance, trunk usage, transaction codes, emergency recording and alerts. In addition to standard reports, systems may permit supervisors or administrators to develop custom reports, unique to their own Call Center requirements. Raw data is stored for some management-defined period of time, and is available for additional reports for a period of at least a week, up to one or more years.

Monitoring

Displays and reports tell only part of the story of agent performance. To really evaluate agent performance and to assess the quality of service provided, supervisors must be able to listen to agents in actual phone conversation with callers/called parties. This has traditionally been accompanied by supervisors walking to the agent's desk and listening to the call by plugging a second headset into the station set. This approach has become an accepted part of Call Center culture for many organizations. However, the same result can be achieved with silent and split/silent monitor.

Silent monitor allows the supervisor to listen, undetected, to both the agent and the caller/called party. If necessary, the supervisor may join the call as an active participant at any time. Split/silent monitor allows the supervisor to listen, undetected to both the agent and caller/called party. If necessary, the supervisor may prompt the agent and remain undetected by the caller/called party. This is particularly useful in training situations, or where threatening or harassing calls are anticipated. Typically silent monitor is invoked on demand, or it may be timed, or it may rotate through the group as calls are terminated.

Forced Answer

When supervisors notice that the number of calls in queue is rising, they may wish to artificially improve the average time to answer (decrease caller time waiting in queue) by forcing agents to answer new calls as soon as prior calls are terminated. This is done by forcing a group-wide override of the wrap/work features.

Move

In the past, supervisors watched Call Center traffic, and then moved agents to the groups that were most active. Today, it is no longer necessary to move the agent to the call. Sophisticated, intelligent systems now bring the calls to the agents. Nevertheless,

Call Center supervisors want to retain the option of moving agents between groups should the need arise. Thus, systems offer an agent Move command that allows agents to be moved even while busy on an incoming/outbound call. At the end of the call, that call's statistics apply to the original group. Then the agent should get his or her next call from the new group.

Messaging

When supervisors walked around the Call Center to monitor agent activity, they could easily stop and speak to any agent at any time. Now that Call Center systems have eliminated the need to actually go to the agent's work station, supervisors must find a new way to "talk" with their agents: hence, supervisor-to-agent messaging. Supervisors can send text messages to agent telephone displays. These messages may be predefined: "Good job!" or "Waiting time down 15 minutes" or may be created as needed. Alternatively, the supervisor may be authorized to send messages using a wall mounted display unit. Unlike display phone messaging, the wall-mounted display message is visible to everyone in the viewing area.

Night Service

The majority of Call Centers, including those that work a 24-hour day, notice a significant difference in traffic between normal office hours and the night shift. If the center is open at all, it may rely more on automated response systems and will have fewer agents in place to handle calls. Predefined inbound call routing schemes must be modified to accommodate the late hours environment.

An intelligent Call Center system shifts to a night service inbound routing scheme based on time of day. Individual supervisors may also invoke night service manually. When invoked, calls may be routed to another inbound routing scheme, to an answering device, another location or callers may just hear ringing.

Automated Data Collection

This is the ability for callers to use touch-tone signals (or the spoken word) for predefined database inquiries such as account status, balance due, driving directions, etc. Automated inquiry requires an interactive voice response (IVR) system which may be integrated with other functions such as Automated Data Collection and/or Voice Messaging (see Chapter 10).

It prompts a caller to respond to specific questions either using the spoken word or touch-tones. Frequently used in after-hours retail catalog sales or to report service problems in a support center, this feature may be referred to as *voice forms*.

Intelligent Queuing

This provides an announcement to the caller stating the estimated time the caller will have to wait before the call is answered. Frequently the anticipated hold time announcement feature is combined with an option to leave a message, rather than wait for an agent. Instead of estimating the waiting time, the caller may be told how many calls are ahead of him.

Request Callback

This feature allows the carrier to request a callback. This option may be provided in conjunction with the "anticipated hold time" announcement feature, or may be implemented as an option during inbound call routing.

Agent Voice Mail

Permits the caller to leave a message rather than waiting to speak with an agent. Depending on the system, the message may be directed to a group of agents or to a specific agent. This option may be presented at one or more times while the caller is in queue holding for an agent. Voice Mail as part of a standalone ACD may not incorporate all of the functions of standard Voice Mail. Be sure you're clear in defining your expectations when purchasing a system.

Fax

Routine requests for information can frequently be handled by a fax server (see Chapter 11). Order confirmation and general information can be provided very quickly with integrated fax response functionality.

These features give the caller an element of control over the way the call is handled. By using automation to eliminate the ambiguity and frustration of long hold times, Call Centers can increase customer satisfaction without increasing staffing levels.

Deflection

The cost of calls to an 800 number mounts rapidly as callers wait for an agent to answer. To decrease the cost of callers waiting in queue for an answer, deflection decreases hold times by having callers reach a busy signal or by routing the calls to another destination. This happens when the system recognizes that a management-defined threshold has been reached. Deflection is typically based on the number of calls in queue. A more effective parameter offered by some systems is "time in queue." In this case, a caller receives a busy signal when the longest call in queue has been in queue over "x" minutes or seconds as defined by management.

Network Services Often Associated With Call Centers

(See Chapter 9 for more information)

There are a variety of network services (outside lines and their capabilities) that can be used to expedite Call Center function. Among these are:

— *DNIS* (Dialed Number Identification Service): Many separate 800 telephone numbers may share an "800" trunk group. DNIS allows the receiving Call Center equipment to recognize the digits the caller dialed, and then to base inbound call routing accordingly. The digits dialed may be translated to a "reason called" and displayed for the agent, prompting an appropriate greeting.

— *ANI* (Automatic Number Identification): In some locations, the telephone number from which the incoming call is placed can be recognized by the Call Center equipment. This can then be used for inbound call routing or for an automated database access.

— *Interflow*: This provides automatic transparent traffic rerouting based on predefined times/days or traffic thresholds. This is useful for multi-site Call Center networks where call volumes and hold times may be high in one location while there is idle time at another location. Or a business may span several time zones and management may want to avoid overtime in one zone by routing calls to another location. Interflow helps Call Centers attain optimum agent productivity while controlling toll costs.

Agent at Home

Just as management may want to send calls to remote Call Centers, they may want to send calls to individual agent homes or small storefront locations. This capability is particularly applicable for accommodating employees with special needs. In addition, it may be useful in disaster recovery situations. The work at home solution should be nearly transparent for both agents and supervisors, meaning they have the same features, regardless of location.

There is more to designing and managing a Call Center than selecting and implementing appropriate features. Consideration must be given to the context of the center:

1. How it relates to other units within the greater enterprise, the local job and labor force.
2. Customer expectations.
3. Competitors' behavior.
4. Vendor relations.

Although we may think of a Call Center as an independent entity, it does not operate in a vacuum. Businesses where the sole function revolves around Call Center activity (example: retail catalog sales) are the exception, not the majority. Increasingly, Call Center functionality is being incorporated into the general business environment (i.e.: service desk, help desk, account inquiries, telemarketing). Thus, the relationship between the Call Center and the other operating units becomes an issue.

Ideally, Call Center personnel have access to all appropriate corporate databases and telecommunications facilities. Networking must be addressed, along with questions of interoperability and administration and management of the various systems in use throughout the enterprise.

The integrated approach to enterprise wide communications, including telephony, data communications, local area networks and wide area network access is appealing to Call Center managers who appreciate the impact of their role within the larger organizational structure. In addition, managers of other operating units may want access to the customer communication center's real time information. Activity levels by group, product or function should be available at any time to any manager who needs it. The customer Call Center must be an integral part of the enterprise network.

Ideally, all communications services within an enterprise will be available to all users. Information can flow freely to and from every desktop. This unified approach enables workers to do their jobs at the same time that it facilitates system administration and management.

The Call Center Approach to Office Telephone Systems

Many business people are beginning to view their entire company or certain departments as Call Centers, and would like them to have similar functions to those found on ACD. They are particularly interested in functions which provide information on the handling of customer calls and on staff members' productivity. These systems will have to be more flexible and collect more information than today's ACDs. They will also have to combine functions such as multiple appearances of extensions for call coverage with the capability to track progress and status of calls as they move through the organization (transferred, conferenced, put on hold, etc.). PBX developers are beginning to focus on this marketplace requirement.

📖 For more information on setting up a Call Center see The Call Center Handbook by Keith Dawson and The Call Center Dictionary by Madeline Bodin and Keith Dawson. Also, subscribe to "Call Center Magazine" (order from 1-800-LIBRARY or www.telecombooks.com).

CHAPTER 6

Voice Mail and Automated Attendant

Voice Mail and Automated Attendant are presented together because they are almost always functions of the same system. They may be collectively referred to as *Voice Processing*. Throughout this chapter, when we say *Voice Mail*, the capability for Automated Attendant is included.

Definition by How it Sounds to the Caller

The easiest way to define the difference between Voice Mail and Automated Attendant is by what you hear.

Here's an example of an *Automated Attendant* working with a PBX:

> "Thank you for calling Telecom Books. If you are calling from a touch-tone phone and know the extension of the person with whom you wish to speak, you may dial it now. To purchase advertising, dial 1; to order a book, dial 2; for a company directory, dial 8; if you are calling from a rotary phone or need assistance, please wait. Someone will be with you shortly."

The Automated Attendant can be likened to a live switchboard operator who directs calls to the appropriate extensions. It can answer and direct multiple calls at the same time as long as there are enough ports (one call uses one port until the call is answered at the extension, freeing up the Automated Attendant port for the next call.) Most Automated Attendant systems can respond only if the caller is using a touch-tone telephone (a few respond to rotary dial signals). Some systems will also recognize the spoken word such as yes, no or saying a number, a name or a phrase. This is called *voice recognition.* Since voice recognition does not work 100% of the time, the system can be set up to transfer a caller to a live person if it does not understand the word spoken.

Voice recognition is proving to be particularly useful in large corporations, especially when callers don't know the spelling of the name of the person they're calling and do not want to listen to a long list of possible options. When trying to use the directory capability of the voice mail, saying the name can be easier. The system can be set up to look for all possible spellings based upon the way the name is pronounced.

Simultaneous answering of calls requires the capability to provide multiple announcement recordings.

Now, if as the caller, you use the Automated Attendant to dial an extension and that extension is not answered, after a few rings you will hear the person's *Voice Mail*:

> "This is Harry Newton on Friday, June 10th. I'm not in my office. Please leave a message at the tone, or dial 0 to reach our switchboard attendant."

You wait for the tone and leave a personal message for Harry Newton which only he can retrieve (by using a password). While you're leaving the message, you are using a port. When Harry retrieves the message, he is using a port. That's *Voice Mail.*

📖 Some of the following information comes from a book entitled, The Voice Mail Reference Manual & Buyers Guide by Marc Robins, Robins Press (718-548-7245).

The first Voice Mail system was installed in 1980 by VMX, Inc., started by Gordon Matthews, the developer of Voice Mail.

VMX has since merged with Octel, which has subsequently been purchased by Lucent Technologies.

In the beginning, there was a lot of resistance to the idea of "having the telephone answered by a machine," but the barriers to acceptance have dropped along with the prices. Voice Mail is here to stay. As you will read in the chapter on Computer Telephony, Voice Mail capabilities are being improved on further using the concept of *Unified Messaging* which enables you to view as well as hear your Voice Mail messages through the local area network (Personal Computers linked together).

Physical Components of Voice Mail and Automated Attendant

The Voice Mail hardware may reside on a shelf in a PBX cabinet or be housed in a separate cabinet or on a PC. It typically works in conjunction with a PBX or Centrex system.

The Automated Attendant and Voice Mail may also work with an Interactive System for Voice or Fax response.

Voice Mail's physical makeup may resemble that of the PBX. There are circuit boards, each having a certain number of ports, usually two, four, or eight. These circuit boards slide into slots on a shelf either inside the PBX cabinet or into a separate cabinet or PC. Voice Mail systems are increasingly PC-based.

The number of ports represents the number of callers who can be using the Voice Mail system simultaneously . There are different types of system memory, controlling the system functions and storing the Voice Mail messages left in user's mailboxes. There is software controlling the system functions which may be accessed by a terminal or PC to make programming changes. As with PBXs, no two systems work in exactly the same way, so nothing that you learn about one system can be assumed to apply to another.

Different Ways of Setting it Up

There are some systems that work only as an Automated Attendant, incorporating no Voice Mail capability or working in conjunction with a separate Voice Mail system. Other systems work only as Voice Mail although most Voice Mail systems may be set up for Automated Attendant as well. Most PC-based systems incorporate both capabilities.

In some older Voice Mail systems certain ports must be designated for Automated Attendant only and others for Voice Mail. More flexible systems use available ports for either Automated Attendant or Voice Mail functions, depending upon the requirements of the moment. These are called *dynamic ports*.

Some systems are set up as Voice Mail only, even though they have the Automated Attendant capability. A PBX may enable callers to directly dial each individual extension (called *DID* or *direct inward dialing*). If an extension is busy or not answered, the caller hears the Voice Mail greeting of the person he called.

We have mentioned that Voice Mail may work with or without a PBX, but most work with one. Some of the major PBX manufacturers, such as Lucent, Siemens and Northern Telecom, sell Voice Mail systems to work with their particular PBXs. Although these can work with other PBXs, they are designed to integrate with the manufacturer's PBX and often have the appearance of being part of the PBX. There are other major players in the Voice Mail market including Centigram, AVT and ActiveVoice; and many smaller companies.

The Voice Mail systems from these other companies are often designed in cooperation with PBX manufacturers and may integrate with the PBXs as smoothly as the Voice Mail systems sold by the PBX manufacturers themselves.

Most of the PC-based systems now have NT operating systems and may have more capabilities at a lower cost, as well as lower upgrade costs than the non-PC-based proprietary systems. Some of the newer "communications server" type telephone systems (see Chapter 12) have even lower costs for Automated Attendant and Voice Mail, since they are an integral part of the system rather than separate systems. The important thing is to understand which functions are possible and which are important to your organization before you decide on a system.

Whether the Voice Mail is made by the PBX manufacturer or by a different company, it is still necessary to have a sufficient number of connections between the Voice Mail system and the PBX. These may be either analog or digital *ports* on the PBX, requiring separate circuit boards.

You may also install Voice Mail if you use *Centrex* telephone service where your telephone system switching and other functions take place back at the central office of the local telephone company. If you use Centrex service and choose to buy a Voice Mail system housed in your office, you must connect it to the telephone company central office with a circuit known as an *SMDI link* (station message detail interface).

The local telephone companies also sell *central office based Voice Mail* which may be rented for a monthly fee along with each Centrex telephone number. In this case, the Voice Mail system resides at the central office and not on your premises.

It may be preferable to have one telephone installation company responsible for both your PBX and your Voice Mail. However, many companies elect to have a separate Voice Mail vendor since they want to take advantage of some of the more advanced features often not available with the Voice Mail systems sold by the PBX manufacturers. This is changing somewhat as the systems sold by PBX manufacturers improve. Another view is that specialization will become the rule and we will buy the PBX from one company and the voice processing systems from another. Newer types of telephone systems will include Voice Mail and not require a separate system.

Using Automated Attendant

Some businesses use the Automated Attendant to answer calls only if the live switchboard attendant is busy on other calls. In this instance, you may want to install the Automated Attendant so that it will only answer calls coming into the outside lines further down in the *hunting group*. For example, the main telephone number, when busy, rolls over to four more lines which are all answered by a live person. If the incoming calls reach the fifth line or beyond, and the switchboard operator does not answer within several rings, then the Automated Attendant will answer.

Another way to use Automated Attendant is to ask callers other than customers (such as employees and vendors) to use a different telephone number which is answered by an Automated Attendant. This keeps the live switchboard attendants free to respond to customers.

For example, a company may have ten incoming lines 883-1000 through 883-1010. When someone dials 883-1000 and the call comes in on one of the first five lines, a live switchboard operator answers. Employees and vendors can be instructed to call 883-1006 which is the sixth line in the 10-line rollover group. Callers to 883-1006 are answered by the Automated Attendant and can direct their own calls, keeping the live switchboard operator free for customer and new business calls.

Voice Mail System Features

Some desirable Voice Mail system features include the following:

— The ability for the caller to listen to a message he's just left, add to it, discard and re-record its or leave it flagged as an urgent message.

— The ability for the caller to escape to the switchboard attendant or another live person at any time by pressing "0."

— The ability for a caller to leave one Voice Mailbox and get into another one at any time by dialing the new extension number.

— The capability to leave a single message, but have it sent to a group of preselected Voice Mailboxes known as a *distribution group.*

— The ability for the Voice Mailbox user to dial in to retrieve not only voice messages, but faxes as well.

— If the system can do *text-to-speech **conversion*** enabling the system to actually read e-mail over the telephone, the ability for the Voice Mailbox user to dial in to retrieve e-mail.

— The ability to dial out, activating the paging beeper of the mailbox user.

— The ability to dial out for the purpose of sending the caller to a customer service person located at another location. May be one selection of an Automated Attendant menu.

— The capability to access a company directory if the caller does not know the extension of the person he is calling when encountering an Automated Attendant.

— The capability for the mailbox user to slow down or speed up a message being reviewed.

— The capability to understand the spoken word so that the Automated Attendant can direct calls using spoken directions. For example: "To reach our sales department press "1" or say "Yes" now.

Voice Mail System Components

Voice Mail system components:

1. A *CPU* (central processing unit), which is the brain of the system and provides the processing power. Many Voice Mail systems are PC-based and use the processor of the PC.

2. As with a PBX, the Voice Mail system uses *software* to provide the intelligence controlling the system features, including the ability to integrate with the PBX.

3. The system includes circuit boards called *codecs* which convert the spoken word into a digital format for storage and then back to an analog voice for retrieval of the message. The rate of speed at which this conversion takes place is dependent upon the sampling techniques used by the manufacturer to convert the analog signal to a digital form.

4. *Hard drives* provide the storage for the Voice Mail messages and are also the repository for the announcements and other recordings inherent in the system. PC-based Voice Mail systems or Telephone Systems with Voice Mail included tend to provide more hours of storage at a lower cost. Some PBX vendor Voice Mail systems offer only 10

hours with a 6 port Voice Mail system, while a PC-based system usually provides 80 to 120 hours as standard.

5. *I/O cards* (input/output cards) are circuit boards providing the physical connection to telephone lines, telephone systems and other related equipment. They also accept the DTMF (touch-tone) signals which the callers use to communicate with the Voice Mail system. This is called *tone detection circuitry.*

The *voice mailbox* itself is an electronic location in the system on which messages are held for each system user. The mailbox owner calls in and retrieves messages using a password entered from a touch-tone telephone. He can listen to messages, save them, erase them or forward them to another person's mailbox.

Voice Mail systems offer a variety of functions both for callers leaving messages and for mailbox owners. They all perform the basic operations of recording messages, receiving messages and redirecting messages. Additional capabilities vary from one system manufacturer to another.

Most Voice Mail systems use menus which are *treed*, meaning that you select a feature from an opening menu and are then branched to another menu, etc. Different systems use different keys on the touch-tone pad to access different features (or, if using voice recognition, different words or numbers).

More Voice Mail System Capabilities

Basic recording features may include the capability:

- To listen to the message you just left before sending it;
- To edit the message you just left before sending it;
- To erase and re-record the message you just left;

- To specify urgent or future delivery of your message;
- To flag your message as private for extra privacy;
- To broadcast your message to a group of mailboxes at the same time.

Basic receiving features may include the capability:

- To listen to your messages;
- To reply to messages from other users of the Voice Mail system automatically;
- To forward a message to another mailbox, appending your own comments at the end;
- To save messages for future reference;
- To delete messages;
- To control message playback by rewinding, backing up incrementally, advancing incrementally, pausing, and/or changing the speed or volume;
- To skip over messages.

The mailbox owner may also have the capability to:

- Change passwords and personalize greetings;
- Toggle among multiple greetings;
- Verify delivery of messages to other mailboxes on the same system;
- Check the status of the mailbox (full, empty, etc.);
- Set up group distribution lists for broadcasting a message to a preselected group of mailboxes.

Voice Mail systems also provide the following features:

- **Message waiting indication**. A Voice Mail system working in conjunction with a PBX can activate a *message waiting lamp* or indicator on the telephone of the mailbox owner. If there is no capability for this, the Voice Mail may periodically ring the telephone to indicate a waiting message or provide a distinctive *stutter dial tone*, heard when the telephone is next used. Some Voice Mail systems can call a beeper number. (You would be beeped to indicate that there is a message in your Voice Mailbox.)
- **Message delivery options**. This enables the mailbox owner to have his messages follow him around. It is possible to program the system to forward messages to different telephone numbers at different times of the day.
- **Guest mailboxes**. You may elect to have individuals outside of your company (guests) have a mailbox in your Voice Mail system (e.g., consultants, subcontractors, attorneys, accountants). The guests can send messages to mailbox owners on the system and receive messages from them. If you plan to use guest mailboxes, make this clear when you are setting up the system. You may need to assign corresponding extension numbers from the PBX.

Voice Mail Administration

Another consideration with a Voice Mail system is its administration. In most instances, there is a separate PC-based administration terminal which may sit next to the PBX administration terminal. Management and administrative functions include: configuring the system for the type of telephone system connections required; setting up specific system functions and applications; setting up mailbox parameters and enabling system features; updating and changing passwords, mailbox numbers and voice prompts; and tracking system usage.

PC-based Voice Mail systems using an NT platform make administration easier if you already have someone who is trained to administer an NT Local Area Network (computer network). Newer telephone systems that integrate Voice Mail within the system have common administration for the telephone system and Voice Mail functions, sometimes using the same type of software as is used for browsing the Internet (called a *browser based interface*).

The Voice Mail administration (like the administration of some telephone systems) can be accessed from any computer on the Local Area Network.

Administrators can dial into the system from a remote location to make changes and run reports.

Administration and management features available on most systems include:

- — Configuration of individual ports for Voice Mail, Automated Attendant or other functions;
- — Connecting individual ports to DID or combination trunks, tie lines or PBX extensions;
- — Setting up company, night and other greetings for the recorded announcements that callers hear;
- — Setting up the number of digits in the mailbox numbers;
- — Setting up the level of security (passwords, passcodes, etc.);
- — Setting up what is called the *class of service* for each mailbox. This is the length and number of messages which can be left before the mailbox is full. *Note*: Newer systems

dynamically allocate the hours of storage, minimizing the likelihood of a "mailbox full" announcement.

— Generating statistical reports on such information as the use of mailboxes, use of the system memory, volume of telephone traffic through the system, and ports in use at different times of day.

Voice Mail System Implementation and Integration with the PBX

One of the most important issues relating to the success of a Voice Mail system implementation is how well it integrates with the PBX . Evidence of poor integration includes such things as:

— Inability of the Voice Mail system to activate message waiting indicators on the telephones;

— Requiring a caller to re-enter the extension number or telephone number dialed from a touch tone telephone in order to enter the Voice Mailbox of the person called;

— A long wait between a request being made of the Automated Attendant and the Voice Mail greeting of the called person being heard.

— A lot of "dead air" heard by the caller (nothing heard at all).

In addition to issues of integration, there are variables relating to implementation. Evidence of poor implementation may include:

— Callers hearing inappropriate announcements such as, "Your call is being forwarded to the switchboard attendant" when there is no switchboard attendant available (for example, during the evening);

— Callers who know which options to select having to wait for recorded announcements to finish prior to entering the numbers;

— The Voice Mail system not responding to the numbers dialed by the caller;

— In some older Voice Mail systems, a longer touch-tone signal in order to respond appropriately (works better if you hold down the touch-tone button a few more seconds);

— Asking callers to dial complex sequences in order to get help, such as ##0 or *T6668#. One of the largest telephone companies in the U.S. expects callers to dial *five* digits plus the pound sign to reach a sales representative. Talk about *sales prevention*!

— Not giving callers an easy escape route to a live person.

A well-implemented system can be used intuitively by people accustomed to using Voice Mail systems. Pressing 0 is the logical thing to do when you want to reach a live attendant. Anything else is confusing.

Integration of the Voice Mail system with the PBX should not be confused with *interfacing*. Any Voice Mail system can interface with virtually any PBX. As we mentioned earlier, a Voice Mail system can actually be connected directly to outside telephone lines if you wish, but without the call processing functions of the PBX, callers can only leave a message and cannot reach a live attendant or be switched to an extension.

Interfacing implies simply that the Voice Mail system can be reached by dialing into the PBX.

Another characteristic of integration is known as *called party identification*. This means that the telephone system is able to send information about the extension back to the Voice Mail system if the extension is busy or not answered. This enables the Voice Mail to give the caller the appropriate message and to send the call to the Voice Mailbox to leave a message.

Some Voice Mail systems integrate with the PBX through a standard RS-232 link. One thing to remember when adding a Voice Mail system to a PBX is that there will be additional hardware and software required for the PBX.

Networking Voice Mail Among Multiple Locations

Just as PBXs in different geographic locations can be connected in a network and appear to work as one system, so Voice Mail systems can be networked. Messages left in a Voice Mail system in New York can end up in a Voice Mailbox in London if the two systems are networked.

Some of the challenges of networking Voice Mail systems include:

- Maintaining the voice quality of messages sent back and forth over the network;
- Providing a method of ensuring that messages are reliably delivered across the network and having notification of messages that are undeliverable for any reason;
- Allowing for flexibility in configuring the network, including the number of PBX network connections, the number of Voice Mail connections and the number of different delivery options;
- Maintaining directories of users and systems for each system on the network so that they are all updated with changes and additions at the same time;
- Providing service and support for the entire network.

Networking is done best with systems that are all from the same manufacturer. Attempting to network Voice Mail systems from different manufacturers may not be practical, since they do not all use the same commands.

There are different ways to accomplish networking multiple Voice Mail systems. If there are no permanent physical circuits connecting the sites, when a message is left in one Voice Mail system it will periodically dial into one of the other Voice Mail systems to transmit the messages to users in that system. The disadvantage of this is that there is a delay from the time the message is left until it is delivered, and there is a charge for the telephone call as one system dials into the other.

If your organization already has a WAN (wide area network) in place with permanent physical connections among sites for connecting the computer network, you may be able to use these circuits to immediately transmit Voice Mail messages left in one location to a Voice Mail system at another location. In this case, the message is sent in a digital form over the circuit and the Voice Message is reassembled at the distant end. This is a more efficient method of Voice Mail networking than having one system dial another at intervals.

As telephone and Voice Mail are now being introduced to the computer networks (LANs and WANs), the capability to network Voice Mail is taking on new dimensions. Read more about this under *unified messaging*, in the chapter on Computer Telephony (Chapter 12).

Voice Mail Service Bureaus

Voice Mail service bureaus are companies who have installed Voice Mail systems on their own premises. They rent individual Voice Mailboxes on these systems to subscribers for a monthly fee. Many of the traditional live telephone answering service companies have changed completely or in part to Voice Mail service bureaus.

Users of the service bureau may call forward their office telephone numbers to their own unique telephone number (usually a Direct Inward Dial number) at the service bureau. Or, if they want all callers to reach the Voice Mail, they may just give out that telephone number as their business number. Service bureaus can also provide individual 800 numbers with a service called *DNIS* (dialed number identification service), which is provided to them by a long distance company.

Many Voice Mail service bureaus offer the same features as an in-house system, including: message recording; reviewing and editing; password security for mailbox owners and various message delivery options, such as dialing out to another telephone number or dialing a beeper.

We have mentioned Unified Messaging several times in this chapter. Refer to Chapter 12 to see how Voice Mail, e-mail and fax messages now use unified messaging so that they may all be viewed on your desktop computer screen and accessed by telephone if you are out of the office.

When you are purchasing a Voice Mail system, here are some questions to ask:

- How many ports? This identifies the number of simultaneous users at exactly the same time.
- How many hours of storage for messages left and for announcements?
- What are the maximum number of ports and hours for the system purchased without having to change the CPU (central processing unit)?
- Is the operating system NT, OS2 or some other proprietary system and what, if anything, does this mean to you? Considerations may be ease of administration (if you know

how to administer NT for your computer network it will be easier to learn to administer an NT-based Voice Mail system) and whether you want to make the Voice Mail system or its administration terminal part of your Local Area Network.

— Is the Voice Mail capable of Unified Messaging using GUI (graphical user interface) and TUI (text to speech user interface)? (See Chapter 12.)

— Who will install, service and support the system, and what is their track record with other customers in your neighborhood? (When things go wrong, it helps to have support nearby.)

The next chapter will discuss two other types of systems which monitor PBX activity – Call Accounting Systems and Facilities Management Systems.

CHAPTER 7
Call Accounting and Facilities Management Systems

Call Accounting

Call Accounting is a tool to help organizations manage and control telecommunications expenses. A Call Accounting System works in conjunction with a PBX or Key system to track incoming and outgoing telephone calls. The resulting detailed and summarized reports provide information on the duration, destination and cost of the call and on the extension number making or receiving it. The information is used to track and charge back expenses to departments and to minimize unauthorized calls.

Most typically, the expenses tracked are outgoing long distance telephone calls. However, an increasing number of or- ganizations are tracking local calls, fax calls, modem calls, Internet access and also incoming calls – especially toll free calls (800, 888, etc.) that are paid for by the recipient. Some organizations track internal calls as a means of determining what is going on within the organization.

Call Accounting systems can also capture the calling telephone number (*Caller ID* or *ANI*) to keep a record of who is calling in. You can also manage other, non-usage based expenses using Call Accounting charge back systems.

Organizations differ in their methods for charging back or allocating costs to departments or individuals. The Call Accounting report costs can represent the actual amount to be charged back or can determine the proportion of all telecommunications expenses that should be assigned to each department. It is important to note that the Call Accounting report never matches the actual bills exactly, due to bills having different start dates and to variables in the costing of individual calls. Call Accounting systems also do not have what is called *answer supervision* (detecting when someone at the other end actually answers the telephone). If someone dials a call and lets it ring for a long time, it may show up on a Call Accounting report as a completed call, even though it will not appear on the bill.

Many organizations are beginning to distribute the reports to individuals using company *e-mail* or the *Internet*, rather than generating the voluminous paper printouts of the past. These new methods of distribution facilitate more frequent reports, often weekly instead of monthly, which keeps information fresh and keeps the pressure on to manage expenses.

These reports can be customized and sent automatically, keeping administrative costs down.

All Call Accounting systems being sold today run on a PC and are Windows-based. The size of the hard drive required depends how many *call records* will be stored (one record = one call) and how often you want to print or distribute reports.

If a company uses *account codes,* this must also be considered in terms of the amount of storage on the hard drive. Account codes are digits dialed in after the telephone number to assign the cost of the call to a specific project or client.

It is a good idea to buy a hard drive that is twice the size you will need to sort your data. In the event of a system problem, you may want to store all of your call records to date. Having the additional disk space will enable you to do so.

Many Call Accounting systems are set up on a separate server on the Local Area Network. It is not recommended that the system share a server or PC used for other functions. If this is the case, the call detail being captured is backed up along with the regular backup of all information on the LAN.

The information coming out of the telephone system into the Call Accounting system is called *SMDR (station message detail recording)* or *CDR (call detail recording)* output. It may also be called *RS-232 output* since it connects to the Call Accounting system using a cable with RS-232 connectors at each end. This comes out of the RS-232 port of the PBX which is also called the serial port. This is not automatic. The PBX must not only have the capability to provide this output (most do), it must also be programmed to do so.

Most telephone systems can work with Call Accounting systems. Some telephone systems are more efficient than others in terms of the amount of disk space required to store the call detail records they send out over the serial port.

The Call Accounting software also includes information on rates (the cost of telephone calls,) which it then applies to each call depending upon the duration, destination and time or day. As mentioned above, the Call Accounting pricing does not exactly match the cost of the call as it is billed by the long distance carrier or local telephone company. It usually comes within five to ten percent if the rates have been kept up-to-date in the system.

Rates may be computed through the use of *V&H (vertical and horizontal)* coordinates incorporated into the software. These pinpoint the geographic location of the destination for the purpose of assigning a cost to each call.

Most organization's long distance pricing is now based upon one of many *tariffs* from the long distance carrier, with some percentage discounted each month from that tariff. Typically, different types of calls have different costs per minute. Under some contracts, costs can change with rate increases. An attempt can be made to have the call accounting rates match the actual rates, but practically speaking, it will be less work to have it approximate rates and to charge back the actual bill amounts based upon the proportion of usage by each person or department in a given month.

The Call Accounting system supplier can provide rates in the system based upon one of the long distance carrier tariffs, or can customize your system based upon your own actual pricing plan. Some organizations do not want the actual cost of calls to appear on the Call Accounting report as, if the rate is very low, it may encourage people to use company telephones for personal calls.

Call Accounting software is hierarchical and can provide information on many different levels. Not all systems provide the same number of levels or the flexibility in setting up report parameters, so it is important to determine what information you require before selecting a system.

Most systems have four levels providing the following reports:

- **Extension**: A list of all telephone calls made by a particular extension.
- **Department**: A list of the calls or summary of costs for each extension within a department.
- **Division**: A summary of the costs for each department in the division.
- **Company**: A summary of the costs for each division within the company.

The dates for the report can be specified. The system also has the capability to store historical data and to provide year-to-date information for each category.

Call Accounting is one of an increasing number of systems peripheral to the telephone system. Each of these peripherals faces the challenge of keeping a database of system users with their correct extensions, fax numbers, cell phone numbers and e-mail addresses up to date. Part of the challenge when implementing a Call Accounting system is determining how this will work.

Some Call Accounting systems may be designed to communicate with a particular type of PBX database, so that when someone changes a name or telephone extension, the Call Accounting database is automatically updated. Ideally, system interfaces are built to create one unified database, but practically speaking, this never happens. Part of the reason for this is that different departments within the organization do not cooperate in a common effort (i.e. Human Resources, Information Technology and Telecommunications).

It is best if the PC designated for the Call Accounting system is not used for any other function since the hard drive must always be available for the collection of calls. Sometimes a *buffer box* collects the call detail output from the PBX and stores it until the buffer is polled by the PC. In a company where there are multiple locations, it is possible to have just one Call Accounting system and buffers (sometimes called *black boxes*) at the remote sites collecting information from the telephone system at that site. The Call Accounting system can call into each of those buffers to collect data. This is called *polling*.

Call Accounting software is sold by a group of companies who specialize in this area. Most are relatively small with annual sales under $2 million and often have their systems concentrated in a particular region. Some companies sell the software directly and others sell through distributors. You may want to consider purchasing the Call Accounting system from a company who specializes, such as Newcastle Communications in New York City (phone #212-780-9680). As when purchasing any other type of system, check the track record of both the company making the Call Accounting system and the company who will install and support it.

Most telephone installation and maintenance companies will sell you a Call Accounting system to work with your telephone system. If you buy a Call Accounting system from them, be sure that a sufficient number of their customers have the same system to ensure that you will receive continuing support.

As the incidence of Toll Fraud increases, Call Accounting systems are taking on new importance. Many are sold with Toll Fraud Packages included or optional.

Toll Fraud refers to the unauthorized use of your telephone system by outsiders to make long distance calls on your lines. Toll Fraud is a multi-billion dollar illegal industry and all organizations are vulnerable. Your telephone system may be "hacked into" through your 800 number, your Voice Mail system or the remote access port used by your telephone installation and maintenance company to diagnose problems.

The Toll Fraud package in your Call Accounting system will not prevent hackers from gaining access, but it will recognize unusual calling patterns through your telephone system which may indicate unauthorized use in progress. Some Call Accounting systems dial a beeper number to immediately alert you to a problem. Others will print a warning on the screen of the system administration terminal or on the paper report. You can also be notified using e-mail.

Although most organizations use Call Accounting to obtain basic reports on who made what call, others are more innovative.

For example, if you advertise in several different publications, when someone calls your company, the switchboard attendant can ask where they saw the ad, then dial a three digit code number relating to that ad before sending the caller through to the salesperson. The resulting report will indicate the volume of calls you are receiving from each ad, helping you to spend your advertising dollars more effectively.

Another function of the Call Accounting system is to track call activity on each of your outside lines, which is known as a *traffic study* or *traffic report*. If you request a traffic study from your telephone installation and maintenance company, there may be a charge of up to $500. If you have a Call Accounting system you may do it yourself as often as you wish.

A traffic study lets you know whether all of your outside lines are in use and if so, how often this happens. If you have 25 outside lines and only 15 are in use at the busiest times of day, you will be able to disconnect lines and lower your monthly costs. If the traffic study shows that all 25 lines are in use at the busiest times of day, this means that your callers are hearing busy signals when they dial your number and you need to order more lines.

Call Accounting reports can also single out calls to a specific telephone number. If you have two offices and you see that you are spending $1,000 monthly to make telephone calls back and forth, you may want to order a *tie line* to connect the offices. This may cost only $400 per month for unlimited usage (one call at a time).

Some systems track internal calls as well, so managers can see how the departments within the company are interacting and who is talking to whom.

There are a few stories relating to Call Accounting. One is about a boss finding out that an employee was calling the boss' house regularly during the day to talk to the boss' spouse. Another is that a company manager had a huge box wheeled through the office with "Call Accounting System" written in big letters on both sides. Even though the box was empty, the company's telephone bill dropped dramatically in the following month.

When employees believe that their calls are being tracked, they think twice before making personal telephone calls. Some companies actually pull reports on each employee's home telephone number to determine the amount of time being spent on personal business.

An alternative to having an on-site Call Accounting system on site is the *Call Accounting Service Bureau.* This type of company polls the buffer device on which your telephone system stores the call detail records. They processes the reports and send them back to you in whatever format and at whatever level of detail you wish. They can also take the information from floppy disks or CD-ROM provided by your local or long distance carrier.

One Service Bureau, Comware Systems, Inc. (phone# 203-326-5500) distributes customized Call Accounting reports to managers via the company *e-mail* system and the *Internet*. This eliminates not only the need for paper, but the time spent looking through extraneous information. Some Service Bureaus enable you to look at your call data on a real time or daily basis by dialing into their site on the Internet.

Hotels are big users of Call Accounting systems. They use them to create your bill for telephone calls when you check out. The system enables hotels to mark up the cost of calls. This is done either by adding a fixed service charge or just by billing a higher cost per minute than the hotel pays to its long distance carrier. In hotels, the Call Accounting may be part of a *property management system*. This uses the telephone system to track other hotel functions such as the occupied and "made up" status of the rooms.

Law firms are also big users of Call Accounting systems. After dialing a telephone number, an account code of varying length may be entered using the touch-tone dial pad. When the Call Accounting report is printed, the calls may be sorted by account code for the purpose of billing the calls back to each of the law firm's clients.

Facilities Management

While Call Accounting focuses the allocation and chargeback of costs associated with calls, outside lines and equipment, Facilities Management systems keep all the information supporting the Call Accounting system up-to-date. Facilities Management is sometimes called Telemanagement. These are broad terms interpreted in different ways by different vendors and user organizations. The remainder of this chapter describes some Facilities Management system capabilities and provides some guidance in purchasing this type of a system.

In 1976, Anthony G. Abbott founded a company called Commercial Software, Inc. which addressed the requirements of managing telecommunications equipment and services using the computer. This was the first company in the facilities management software business. Mr. Abbott is now CEO of Comware Systems, Inc. (203-326-5500), which has a large installed base of telecommunications management software users. The older customers continue to be supported while Comware builds new systems that support the emerging client/server model of computing and telecommunications.

An article authored by Mr. Abbott for *Business Communications Review Magazine* explains Telecommunications Facilities Management:

The primary reasons for using facilities management software are:

1. The responsibility for recordkeeping has fallen on the user. You cannot rely on the local telephone company or the telephone system vendor.
2. As telecommunications expenses increase, the need to manage and allocate these costs becomes even more important.
3. The proliferation of suppliers results in more invoices and separate points of contact.

Facilities Management systems have developed the following applications:

Order/Inventory System

This is a system that automates the process of creating and managing work orders for telecommunications equipment and services. It can reduce the labor associated with creating and managing work orders and, as a by-product, automatically create a detailed inventory of all telecommunications system components. This also interacts with the company directory which is updated automatically as moves and changes are made in the system.

Cost Accounting/Allocation/Chargeback

This is usually the first application that is implemented. Cost allocation has the objective of providing cost center management with meaningful information about the telecommunications expenses, facilitating the exercise of local management prerogatives to control expenses.

This is accomplished by (1) providing detailed reports to upper management, (2) summarizing telecommunication costs for all levels of management and (3) providing journal entries or financial summary information as input to existing corporate financial control systems, such as the general ledger. This permits a review of these expenses in conjunction with normal planning and budgeting activities.

Directory Systems

Directory systems have the objective of maintaining and making available accurate telephone and *e-mail* directory information. They support three directory functions: directory assistance, directory publishing and directory updating. The assistance and publishing functions are generally straightforward. The most complex part is the updating and maintenance activity. Information for the directory comes not only from telecommunications, but also from the Human Resources and Information Technology Departments.

Network Engineering Systems

Changes in tariffs and the continual introduction of new services with varying costs create the need to engineer and re-engineer your communications facilities (circuits) for both voice and data communications, including *Internet* access. Network Engineering support systems fall into two categories: data reduction and simulation. Data reduction is a method whereby detailed usage information is processed and reduced to summary form. Reports that reflect the pattern of calls (e.g., all calls to a particular area code and exchange) and reports that summarize the utilization of existing facilities (traffic studies) are the type of information produced by data reduction systems. Simulation/modeling systems make design recommendations based upon input of detailed telephone usage information.

Trouble Reporting

This application tracks outages, the response time to those outages and a variety of pertinent statistical data which can prove valuable in terms of evaluating how well your telecommunications vendors are doing.

Financial Applications

The financial alternatives offered by the various telecommunications suppliers have created a need for sophisticated financial management tools. Applications such as financial analyses that provide the capability to look at disparate proposals, lease versus lease/purchase and the returns on investment are a necessity, particularly for the large user.

In addition, basic functions such as accounts payable and bill reconciliation add support to the telecommunications management responsibilities.

Cable Inventory

Cable records can quickly become outdated. Failure to keep track of cables and pairs can result in increased expenses in terms of how long it takes technicians to complete work installing new telephones or other telecommunications devices.

The key to a successful computer-based telecommunications management system is the integration of each of its parts so that updating is automatic. For example, a directory system requires information that relates telephone numbers to organizational entities while a usage cost accounting system needs the same information in order to produce its reports. If this information is independently updated, data inevitably do not agree and there are inconsistencies between directories and cost accounting.

Ad Hoc Query Systems

Users of large systems have a variety of information needs which may require answers to questions upon demand such as, "What effect will an eight percent increase in access charges have on my budget?"

Another perspective on facilities management is presented in an article from TELECONNECT Magazine *by Tracey Tucker.* This discusses facilities management software, also known as *telemanagement software.* This includes some tips for purchasing this type of system.

Characteristics Of A Good Telemanagement System

As telecommunications networks become larger and more complex, telecom managers are realizing the importance of controlling every aspect of their network's operations. It's not enough to just track and cost calls. There is a lot of money invested in equipment, and recurring expenses associated with *MACs* (moves, adds, changes). There are maintenance issues to deal with and choices regarding services and equipment. This is why you need tools to help to make timely and informed decisions.

Telemanagement software provides these tools. It automates management functions in three primary administrative areas:

- **Process control**: This refers to the ongoing activities that occur on a day-to-day basis, like traffic engineering, network optimization, trouble tracking and work-order management. These activities are time-consuming and a great deal of information crosses different files. For example, every time you generate a work order, you need to access cable, directory, inventory and vendor contract information. Telemanagement software helps by automating the process and by providing a central database from which you can access all network information.

- **Asset management**: This relates to inventory control. You know what equipment you have, where you have it, and the status of equipment (whether it's already in place, in stock, out on repair, etc.) Asset management also lets you allocate costs for budgeting and track *feature usage* for system design and planning.

- **Resource management**: This allows you to manage your workforce and schedule problem resolution. You can issue work orders based on whom you have available (i.e. Do we use our in-house staff for this job, or the vendor representative?) You are also able to store information on the level of expertise required to resolve a particular problem, so that you can assign an appropriately-skilled technician to the job.

With all the benefits that telemanagement software provides, more organizations are implementing it. Some major differentiating factors are cost, functions and capabilities, and the platforms on which they run. Some offer one or two applications, while others cover the whole spectrum. How do you decide what software is right for you?

1. **Functional Integration.** All of the applications provided by the software should be integrated so that you have one main information repository or database. A system that's integrated in this way ensures that any modifications you make in one module will automatically be updated in all relevant files. Otherwise, you will have a bunch of separate databases for each function which must be accessed and updated separately. For example, each time you issue a work order or trouble ticket, instead of being able to immediately pull up a pair assignment, you will have to go to another database.

   There are a lot of stand-alone systems out there that provide only one or two functions. It is important for them to integrate with a larger telemanagement system. Now you may only have a need for cable and wire management, but later you may want to branch out into network optimization and traffic engineering. If the stand-alone system cannot be integrated, its usefulness is limited.

2. **Product Enhancements.** Ask the vendor to provide a list of the types of product enhancements created over the past few years. What new applications, features and interfaces have been developed? Regular version releases and upgrades are positive signs. The sophistication of the product and the number of applications it provides dictate the enhancement schedule. With a really dynamic product, the vendor may come out with something new every six months.

3. **Portable/Scalable Platforms.** Look for a vendor that provides software applications on a number of different platforms, or at least on the platform to which your organization has

committed. Every organization faces a continually changing business environment. What you have as a computer platform today may not be what you have when your company has been reorganized or merges with another division tomorrow. You should be able to migrate to a larger system when needed.

Also, make sure that this portability is demonstrated. A lot of vendors say that their software can run on different platforms. It often turns out that the software only runs on one platform. For a fee, they will convert the program. To avoid being the "beta site," ask for certification, documentation and references indicating that the software does indeed run on all the platforms indicated.

Documentation is also important. When you switch to another platform, you may not get the same set of functions. Check. When you ask for a demonstration, make sure that it's done on a working system.

4. **Interfaces To Network Management Systems**. Network management systems are used for real-time monitoring and controlling of a telecommunications network. Network management consists of five main applications, as defined by the Open Systems Interconnection or OSI model (a group of standards for communication between computer systems made by different vendors). The applications address management of network faults, configuration, performance, security and accounting.

   Network management takes into account all the smart devices you have on a network, such as PBX's, servers, multiplexers and data switches. This equipment is constantly spewing out management information regarding alarms, traffic statistics, security audit trails, etc. Considering that information, the network management system diagnoses the current situation and decides whether or not to perform network reconfiguration or produce a trouble ticket. When a network management system and telemanagement software are integrated, you have one main information repository. Anything that happens on the network is automatically updated in the administrative database.

5. **Commitment to Open Standards**. Question the vendors' awareness and commitment to open standards. Are they keeping up with specifications and recommendations from the standards committees?

6. **A Diverse Range of Vendor Services**. To what extent is the vendor prepared to support you through product implementation and beyond? Can they provide a turnkey installation? If so, they probably have a better understanding of their platform and the applications they have developed to run on it. This is a definite asset when you enlist the vendor's help to solve any problems that arise.

   Find out also if the vendor provides customization services and, along with this, the composition of the vendor's organization. Look for a company that has a significant group of people in product development, application development and technical support. You will need people that can work with you to help define your requirements and then translate those needs into an actual application.

   The staff's level of expertise is also important. Develop an understanding of their backgrounds. Software is tied very closely to the company from which it comes, so you should assess not only the product, but the people. Get some indication of their knowledge, such as number of years with the company or experience in telecommunications. Also check the rate of turnover in the product development area. Since software is always changing (new features and applications added), if there is a high rate of turnover you may not get a consistent product.

7. **Vendor Stability**. The telemanagement market is a volatile environment. Check how many times a vendor has been bought and sold, changed presidents or modified company philosophy. The vendor's longevity in the industry will affect the product. You will need a close tie to your vendor in order to expand or modify your system in the future. Therefore, choose a vendor with whom you'll feel comfortable for the long term.

   In addition to the vendor's track record, check the product's as well. In some cases, the software has not been developed by that company, but was purchased from the developer. Find out how many companies it has gone through and how many versions there have been.

8. **User Group Activity**. One of the biggest pluses for a telemanagement vendor is that they have a user group that meets annually. User groups put the users in communication with each other. They pool knowledge, share experiences and influence product development. These meetings also give users added clout, ensuring that the vendor continues to support them in a satisfactory manner.

9. **Implementation Support**. Most of the cost and effort associated with a telemanagement system comes from the initial setup of the system. Information that goes into the database must be collected, and a standard coding scheme (i.e., numbers that distinguish between telephones and PCs, for example) must be developed. If you do not already have good records, this process can be labor intensive. Find out if there are any features that facilitate implementation. Autoload capabilities, for instance, enable a PBX to load all its programming information into the system.

Purchasing a telemanagement software system is a substantial investment, so make an informed decision. The above guidelines will give you a start on the selection process.

PART III

Transmission

CHAPTER 8
Cable

This chapter is about the physical transmission medium called *cable*, which is fundamental to telecommunications systems and services. The most well-thought-out telecommunication system will not work as expected unless the cabling infrastructure is properly designed and installed.

There are many variables in both the types of cable and the quality of the installation. Transmission qualities are continually improving as the cabling manufacturers keep up with the demands of the marketplace for sending more information at faster speeds.

Copper is by far the most commonly used type of cable in telecommunications installations. Other than silver and gold, copper is the best metallic conductor of electricity. Silver and gold have other limitations making them unsuitable. Telecommunications signals are *low-voltage* electricity.

Unshielded Twisted Pair Copper Cable

The most prevalent type of copper cable is UTP, which stands for *Unshielded Twisted Pair.* The wires come in pairs which are twisted around each other. The twists minimize the effect of

Figure 8.1

A Twisted-Pair Cable

Compliments of Understanding Telephone Electronics by Steven S. Bigelow

interference (also called *crosstalk*) from external sources, and therefore improve the likelihood that the telecommunications signal will make it through the cable in an undistorted form. This outside interference is *EMI (Electro Magnetic Interference)* or *RFI (Radio Frequency Interference).*

The UTP is rated in terms of the number of twists per inch, more twists being better. If it has at least two twists per inch, it will support not only voice communications, but signals for both *ethernet* and *4 mbps token ring* local area networks for computer-to-computer communications. Two twists per inch is Category 3 cable which will support 4 mbps token ring, but not 16 mbps token ring which requires Category 5.

More twists per foot are better in terms of enhancing the cable's ability to resist the effect of induced (coming from the outside) voltages. Different pairs of wires within the cable are twisted varying amounts.

There are different levels of cable that are UL (Underwriters Laboratories) specified, Categories 1 through 5. The term *CAT 5* or *Category 5* may also refer to a device such as a jack or patch panel, capable of supporting Category 5 cable performance. You may also hear the term Level 5 meaning the same thing.

Category 5 cable has from 16 to 20+ twists per foot with a variable rate of twist. It can support transmission speeds from 20 mbps (megabits per second) up to 155 mbps on a single pair. In comparison, a digital voice transmission is 64 kbps (kilobits per second). A megabit is 1,000 kilobits. Lucent Technologies (formerly AT&T) sells a Category 5 cable known as 2061D. They also sell 2061D plus. Other manufacturers may just call it *plus cable*. It carries speeds up to 620 mbps driving all 4 pairs in the cable.

The copper cable manufacturers must improve transmission speeds to compete with fiber optic cable. 2071 is now available. The speeds which can be supported are dependent upon both the category of cable used and the distance which the signals must travel.

The Concept of Speed or Bandwidth

All signals travel over copper cable at approximately the same speed (60-80% of the speed of light in a vacuum). *Speed*, as it is used in this context refers to **how much** can be sent over a cable or a circuit within a second. So higher speed cables allow much more information (more *bits*) to be sent in one second than lower speed cables. Because more information is being sent in one second, the information appears to be traveling faster, but it is just that more is getting through. Another term for this speed is *bandwidth*.

UTP Configurations

UTP is manufactured in a variety of configurations. You can buy a cable containing one pair or two, four, eight, twelve, twenty-five, fifty, up to 4,800 pair! The most frequently used cable, which is a standard, is a *4-pair* cable (total of 8 wires). By definition, Category 5 is a 4-pair cable. Some 25-pair cable meeting Category 5 standards is also available. Some manufacturers give their own name (or number) to different types of cable.

Each pair of wires within a cable is color coded. For example, one wire is blue with white stripes. It is twisted around a second wire which is white with blue stripes = white/blue-blue/white. There is also white/orange-orange/white, white/green-green/white, white/brown-brown/white, etc.

In many installations, more than one 4-pair cable is run, perhaps two 4-pair cables or, most commonly, four 4-pair cables.

Some forethought is required in terms of how the cable is to be used, although the actual use may be different from the planned use.

Jacks

In addition to deciding how much and what type of cable will be run, it is important to consider how the cable will be terminated at each end. At the desktop, the end of the cable will terminate in a jack into which a connector will be plugged to connect the telephone or other communications device. Sometimes the connector is referred to as the *jack*, but the jack is actually the outlet.

The means for terminating the cable, and the size and configuration of the jack are dependent upon what is to be plugged into the jack. This is not always known when the cable is run.

For example, it is possible to terminate all eight wires of a 4-pair UTP in an 8-pin modular jack. The Category 5 standard requires that you do so. It is also possible to terminate only six of the eight wires (or fewer), leaving one or more spare pairs in the wall behind the jack. This is called a split cable. The spare pairs can also be terminated on a second jack. The connectors plug into the jack, ideally having the same amount of wires as are terminated on the jack. It is possible to plug a 6-pin connector into an 8-pin jack, but it is not recommended in terms of ensuring the best transmission.

The 8-pin connector is larger than the 6-pin. There is also a 4-pin connector that is the same size as the 6-pin connector, but only four wires are terminated on it.

The jack for the telephone in your home is probably a *42A block* (if the telephone is hardwired into it) or an *RJ11C* if the telephone plugs into it with a modular connector (6 pin).

The RJ11C is a little beige square plastic box into which a 6 pin modular connector is plugged (the type of plug at the end of the mounting cord on a single-line telephone). This connects the telephone to the wires, which in turn connect you through the cable in your home, out to the street, and back to the central office of the local telephone company. There your call goes through switching equipment (the *central office switch*) to be connected to the telephone number you are calling. Your voice then travels over the cable and back out to the telephone at the other end. You may hear the term *quad* used to refer to an older type of 2-pair cable. It was not twisted and is generally not used in new installations (quad uses 4 wires, one pair red/green and the other yellow/black).

In what was formerly The Bell System, the division known as *Western Electric* developed some standard wiring configurations for jacks used today. The two you most commonly encounter are *RJ11C* (6 pin modular jack) and *RJ45* (8 pin modular jack). RJ stands for *registered jack.*

Many offices run four separate Category 5 cables to the desktop, terminating each in a separate jack (Figure 8.2). One of these jacks may be used to plug in the telephone. Another may be used to plug the computer into the *LAN* (local area network). A third may be used to plug the computer into a separate outside line, bypassing the PBX or an extension from the PBX, which can be used by the computer. The fourth may be used for a fax machine or some other device.

There is a lot to be said for running more cable to each desktop than you think you will need. Some applications use all four pairs in the cable for one high speed transmission path. It is also important to make the cable accessible. Having to remove the jack

Figure 8.2

Faceplate With 4 Jack Openings

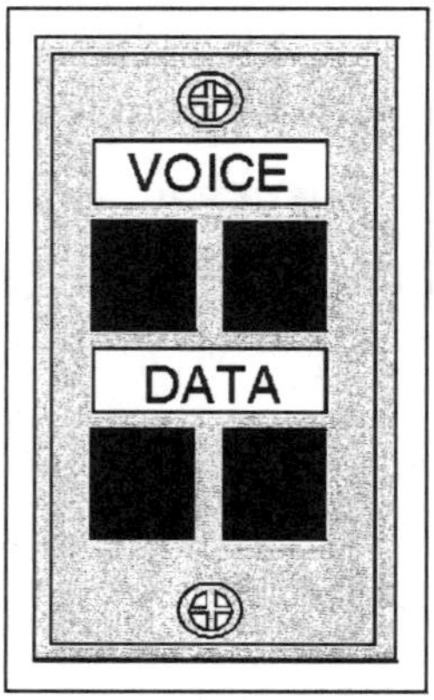

and reach behind the wall to pull out another pair of cables can create a mess, be cumbersome to perform, and may disrupt other devices already working at that location.

Newer types of telephone systems, sometimes called *communications servers*, may result in a rethinking of the amount of cabling to be installed in the office. For some of these systems, the telephone instrument plugs into the LAN connection and the desktop computer plugs into the telephone, so a single set of wires in the cable and single outlet are used for both the telephone system and LAN. In this case, the *communications server*, sometimes called the *telephony server,* is considered to be a node on the local area network, rather than a separate system.

Electric Power at the Desk

Another related consideration is the number of electrical outlets at each desk. Make sure to have sufficient number to accommodate everything that requires power. Most of today's telephones do not need to be plugged in at the desktop. Some power can be sent from the telephone equipment room over a separate pair of wires (*low voltage*) going into the same jack as the telephone.

Certain telephones may require an electrical outlet at the desktop, so it is always a good idea to ask if it is necessary. For example, modules used to expand the number of extensions appearing on the telephone may require power at the desk. Some speakerphones enabling hands-free conversation also require separate electrical outlets.

The newer telephones that plug into the LAN outlet and work with a communications server have separate power cords for plugging in at the desk. The low voltage power that is sent over most telephone cables cannot be sent if the same wires are carrying LAN signals. Whether or not this type of technology replaces the more traditional method where the LAN and telephony signals travel on separate wires remains to be seen.

Main and Intermediate Distribution Frames

We have talked about UTP copper cable and how it is terminated at the desktop. What happens at the other end of that cable?

If all cables in the installation are *home run,* that means that they all pull back to the same place, which is the *MDF (Main Distribution Frame)*. It is a good idea to locate this centrally so that each cable run will be equidistant from the MDF. For more efficient cable distribution, *IDF's (Intermediate Distribution Frames)* are sometimes used. On multi-floor installations there may be one or more IDFs per floor. The IDF is connected to the MDF with a *feeder cable* containing enough pairs of wires to accommodate all of the cable pulls back to the IDF, plus additional pairs of wires for growth. If your installation requires a certain performance of the cables, such as Category 5, be sure that the feeder cables and all associated hardware also support that performance.

The IDFs and MDFs are laid out to accommodate growth and labelled so that records can be kept on the originating and terminating end of each cable. If good records are not kept, you may find it costs less to run new cable than to figure out the old.

The IDF and MDF are made up of *connecting blocks* which are devices made of plastic and metal designed for terminating cable. Although *M66 blocks* are still in place in most installations, newer installations use *110 blocks* developed by AT&T, now Lucent Technologies (Figure 8.3). These are higher in density, enabling more pairs of wires to be terminated in a smaller space. These blocks can be mounted on plywood backboards or can be installed on free standing racks. Sometimes there are *cable trays* above the racks to neatly hold and support the cable.

Some telephone systems are installed using modular *patch panels* at the MDF. This appears to make things easier since anyone can make changes without knowing how to *punch down* (terminate) cable on the block. However, if there is a lot of moving and changing, the modular panel can become a tangle of wires that is problematic to manage.

Another possibility for the MDF which is costly and used primarily for data installations is an *electronic digital patch bay* in which the cables remain in the same place. The connections are changed using software. This method may be used for voice communications in the future. The problem with it is that it creates a single point of failure for the entire cabling system. Hardwired methods of terminating cables are more fault tolerant.

Figure 8.3

AT&T Developed 110 Block

A 600 Pair Block for Main Distribution Frame

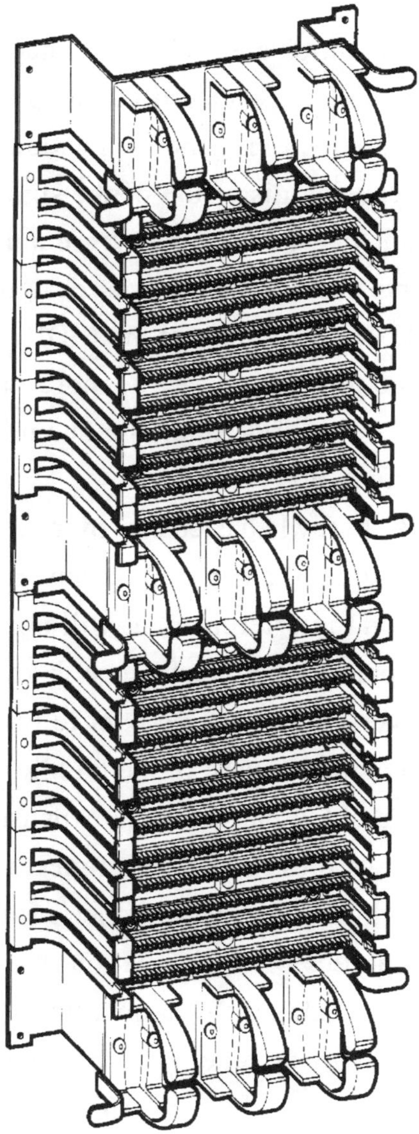

Compliments of MOD-TAP

Shielded Twisted Pair

There is another type of twisted pair cable known as *STP (Shielded Twisted Pair)*. The shielding is an extra layer of metal insulation surrounding the twisted pair. This is sometimes recommended in cases where a lot of outside electromagnetic interference is present or anticipated. If it is not necessary, it is better to do without the shielding, since it can also have the effect of keeping electromagnetic fields generated within the cable trapped inside the shield, causing distortion of the signal. If shielding is used it is important that it be properly grounded. Otherwise, it may attract the interference that it was designed to repel. The grounding should only take place at the MDF (main distribution frame). If it is grounded at both ends, more interference in the signal may result. In general, shielding should only be used in low-level signal situations.

Shielding works best when each pair of wires is shielded separately. Overall shielding of a cable containing multiple pairs will still keep out external interference but will not prevent interference between the pairs. This interference is known as *crosstalk* and results in conversations spilling over to other pairs of wires so that one conversation is heard in the middle of another.

Good quality shielded pair cable is expensive. The expense may be justified since these cables must meet rigid manufacturing specifications which address the diameter and strength of each conductor, the properties of the insulation, the twisting of each pair, the shielding of each pair by a metallic foil, and the shielding of the entire cable by an outer layer of shielding material.

As mentioned earlier, health concerns about the proximity of large amounts of active cabling are creating more interest in the development of shielded cable.

Screened Twisted Pair Cable

Screened Twisted Pair Cable is yet another variation of shielded twisted pair cable developed by ITT. It can carry a signal 75% further than regular UTP in a Category 5 environment.

Grounding

An important consideration relating to the installation of equipment and cabling is grounding. A ground is a common electrical reference point which serves two primary purposes. The first, a *power ground*, is a matter of safety. Metal cabinets and racks housing equipment and cabling should be properly grounded to minimize the possibility of electrical shock to anyone coming into contact with them. This is done through grounding to cold water pipes or building steel.

The second, a *signal ground*, is to ground the cable in order to enable proper transmission and reception of the signal with a minimum of distortion. As mentioned above, in many buildings, the cold water pipes are used for grounding. You may also ground to your *UPS (uninterruptible power supply or battery backup)* Grounding is really "completing the electrical circuit." A complete circuit ensures high-quality transmission. On the matter of safety, however, a person does not want to inadvertently become the object that completes the circuit!

Advantages and Disadvantages of Twisted Pair Copper Cabling

Some advantages of twisted pair cabling include:

- Many developments of new applications call for twisted pair.
- It is a flexible system.

- Distribution cables (feeder cables) are easily used and it is not necessary to home run all cables.
- The same cabling can be used for voice and data communications, saving on materials and labor, and therefore on expense.
- Many technicians are available for performing high-quality installations.
- It is often already in place, so it may make sense to reuse it.

Disadvantages of twisted pair include the following:

- It has limitations in terms of the speed (bandwidth) with which data can be carried, although these speeds are increasing as new cable is developed.
- It is more likely to have crosstalk interference than other types of cable.
- It may be less secure than coax or fiber cable in terms of someone being able to tap into it.

Coaxial Cable

Another type of cable, typically used for data communications rather than voice, is *coaxial cable*, also known as *coax*. It is made up of a center conductor and an outer shielding conductor (Figure 8.4). The center core can be a single solid wire or stranded wire.

The center conductor of the coaxial cable is surrounded by an insulating material called a *dielectric*, which in turn is surrounded by an outer metallic shield. This shield serves both as a return path for the signal and as a shield for the center conductor against electromagnetic interference and crosstalk.

Figure 8.4

Coaxial Cable

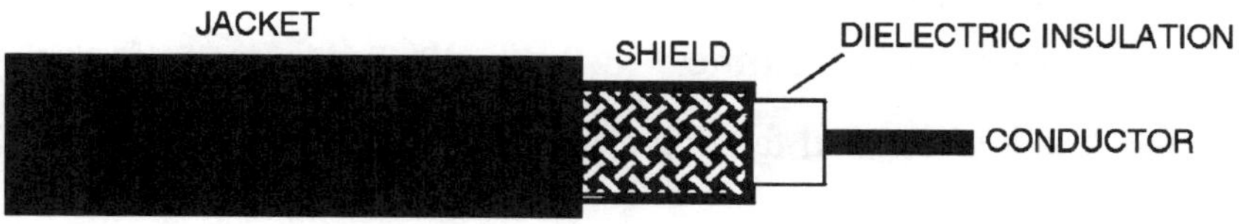

Courtesy of Hubbell Premise Wiring, Inc.

Coaxial cable is used for closed circuit and cable television, video security systems, computer networks and some types of communications systems. Cable television companies are now beginning to use the cable already in place at subscriber's premises to offer telephone service for both voice and data communications. You can have access to 175 channels of video through single coax cable.

Advantages of coaxial cabling are as follows:

- Low susceptibility to electromagnetic interference resulting in less interference and crosstalk.
- High bandwidth for transmitted signals, with resulting lower signal distortion.
- Can be used over longer distances than twisted-pair cables.
- Can be matched to operate with twisted pair through the use of a device called a *balun*.
- Lower signal distortion.
- More channels can be transmitted over the same cable than with twisted pair.
- There is less of a tendency to have crosstalk between cables than with twisted pair.
- There is greater security of the information than with twisted pair.

Disadvantages of coaxial cable are as follows:

- It is more difficult to install than twisted pair, costing more and taking more time.
- It is heavier and fatter than either twisted pair or fiber optic cable.
- Many systems have shifted from coaxial cable to twisted pair as new technology is developed to improve the transmission on the twisted pair.
- It must be installed in either a daisy chain fashion or home run, so it does not have the flexibility of twisted pair.

Fiber Optic Cable

Fiber optics is a technology in which light is used to transport information from one point to another. Fiber optics are thin filaments of glass through which light beams are transmitted. (See Figure 8.5 and Figure 8.6)

Single mode fiber optic cable is used under the streets to connect one telephone company central office to another. It uses lasers as a light source and can transmit a signal for 100 miles without a boost.

Multimode fiber optic cable has a much larger core diameter than single mode. It is typically used within a building, uses LED's as a light source and transmits signals much shorter distances than single mode.

There are two types of multimode fiber optic cable. One is *multimode step index*. This is sufficient to handle the speed of ethernet and token ring LAN transmissions. *Multimode graded index* increases speed 400% over step index. The graded index cable varies the speed of different light paths in the core. This cuts down on pulse widening (distortion) also called *modal dispersion*.

Figure 8.5

The structure of the fiber determines how light propagates through it.

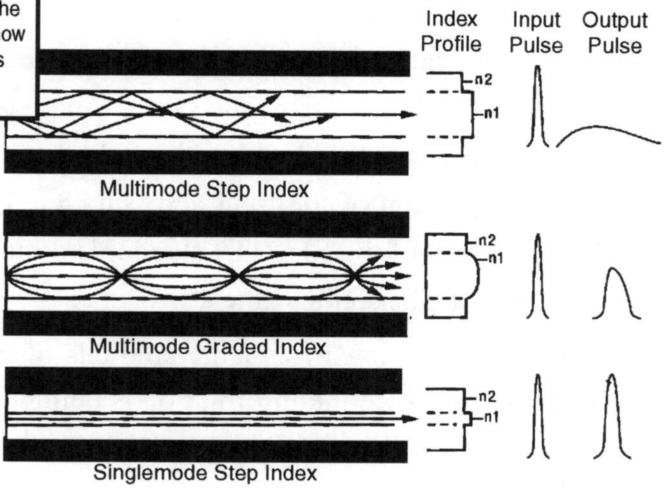

Compliments of AMP and Jim Loizides of The Hylan Group

Figure 8.6

The concentric layers of an optical fiber include the light-carrying core, the cladding, and the protective buffer.

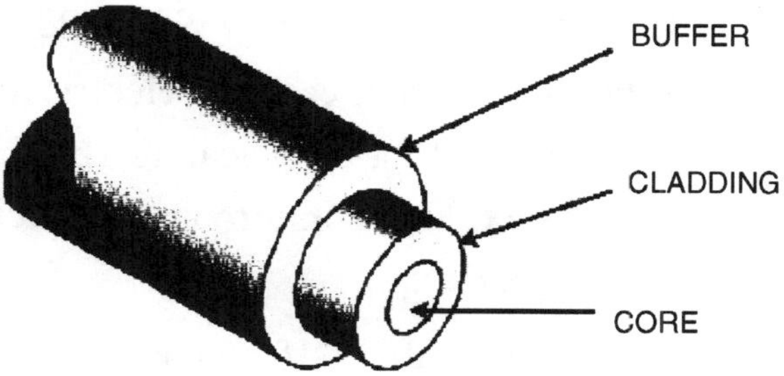

Compliments of AMP and Jim Loizides of The Hylan Group

Fiber optic cable is often used as a *backbone cable* such as a building riser cable. It may also be used between Local Area Networks. It is seldom run to each desktop.

In fiber optic cable design, the trend is towards a narrower core while the cladding remains the same. Coupled with the use of new breeds of lasers, this enables less light to get through, but there is less dispersion of the light and it travels further. These capabilities are part of the EIA/TIA standards definitions.

Advantages of fiber optic cable are:

— Fiber optic cable carries large amounts of data at a very high speed. Since the information is being transmitted by light energy rather than electrical energy, the signal is not subject to the electrical properties of metal wire such as resistance and inductance which can attenuate the signal.

— Signal loss is significantly lower than for either coaxial cable or twisted pair.

— The signal is unaffected by electromagnetic frequency interference. Fiber cables do not produce any electromagnetic noise, nor are they affected by it. There is no need for concern about crosstalk, echoing or static. Fiber optic cable eliminates all grounding problems since it uses optical coupling instead of electrical coupling.

— Fiber optic cables carry no electrical energy so there is no possibility of an electrical spark. Thus, the cable can be used in explosive environments such as chemical plants and refineries.

— Fiber optic cable is far more secure than twisted pair or coaxial cable. There is no electromagnetic field. It is very difficult to tap into a fiber cable without detection, but it can be done.

— Fiber cable is adaptable to almost any LAN configuration.

Disadvantages of fiber optic cabling are as follows:

— There are not very many qualified technicians to install it.

— It tends to be the highest in cost – but this depends in part on the application. The cost is coming down.

Selecting and Purchasing Cable and Installation

When deciding which of the three types of cable to use, it is important to consider the costs of installation, maintenance and future requirements. You must evaluate the suitability of the cable for supporting your applications.

When you are in the process of specifying cabling for the purpose of obtaining pricing, here are some things to ask:

— Will all work be done in compliance with national, state and local codes including building codes, fire codes, electrical codes and health codes? Most areas of the U.S. no longer permit the installation of PVC (polyvinyl chloride) coated cabling in office environments, particularly running through return air plenum ceilings. This coating emits toxic gases when it burns, so it presents a danger in the event of a fire. There is actually a lot of it installed, but the idea is not to put any more in. Teflon or other coatings tend to be higher in cost, but are less toxic when burning. They are now used instead of PVC.

— What is the cabling company's prior experience with this type of installation?

— Will the cabling company be responsible for he replacement of floor tiles, ceiling tiles and any other areas damaged due to the cabling activity?

— What type of cable documentation will be provided at the conclusion of the work? Make sure it is detailed and accurate. Decide how it will be kept up to date.

Building Cabling Systems

Now that we have introduced some specific types of cable used in typical business installations, here is a view on the importance of the building cabling system.

Some building owners or business users of office space view the wiring as a strategic investment. The cabling network must meet both near and long term needs. In the past, the cable design has accommodated the needs of a specific system or systems. With technology changing regularly, the cabling systems being installed now must be much more flexible and universal, capable of supporting all of the communications of a building including voice, data, security, images and video.

The cabling being installed now will be expected to last at least ten to fifteen years, if not more, yet the technology of the equipment using the cable often has a life cycle that is measured in months. Transmission speeds are continually increasing.

This points to the importance of installing what is sometimes called a *structured wiring system*. Due to the expense, not all building owners are interested in installing a structured wiring system for the entire building unless the building is occupied and owned by a single tenant. In multi-tenant buildings, different companies have different ideas about what their cabling system should be. Strictly speaking, the structured wiring system addresses the needs for connectivity from the entrance to the building through the workstation/desktop.

The *riser cable* system, also called the *backbone*, extends vertically from the main distribution frame for the building (usually in the basement). It provides service to the *telephone closet* on each floor of the building. The riser cable may be copper, fiber optic or a combination of the two. In a large, spread out building such as a factory, the equivalent of a riser cable may be placed horizontally.

A pre-determined number of cable pairs are terminated on each floor. These will be used to bring telephone service from the outside world into the space of the company occupying that floor.

The copper riser cable typically uses 24 *gauge* cable formed into *binder groups* of 25 pairs each. The groups are identified by distinctively colored binders and are assembled to form a single compact core covered by a protective sheath. The gauge of the cable relates to the transmission properties and the diameter. The higher the number, the smaller the diameter. Most telephone cable is either 24 or 22 gauge. You may also see the term *AWG* which stands for American Wire Gauge. Twenty-four gauge is typically used for premise cabling, while 22 gauge is used to connect telephone company central offices (although fiber optic cable now connects most central offices).

The *telephone closet* in which the riser cable is terminated is usually located in some common building area, perhaps adjacent to the elevators. It is typical to take the outside lines from the building telephone closet and to run them through a cable in a *conduit* (pipe) into the telephone equipment room. There, the PBX and perhaps the file servers for the LAN, are located. This PBX room is part of the individual business premise.

The plastic and metal blocks on the wall of the telephone equipment room on which the outside lines are terminated are referred to as the *demarc*. This is the point of demarcation where the outside lines are dropped off. From that point the telephone system vendor runs cable to connect the outside lines into the telephone system so that when someone at an extension dials 9 they will be able to place an outside call. Incoming calls also travel this route.

If the backbone is to support transmissions higher than 100 mbps for other than a short distance, the riser cable should also incorporate fiber optic cable. There is a lot more to know about cabling. We have only presented some fundamentals.

Cabling Standards

There are different standards in use that specify characteristics of cable and associated hardware and the way that the cable is terminated. American Standards in use in the United States are established by ANSI/EIA/TIA. This stands for American National Standards Institute/Electronics Industry Association/Telecommunications Industry Association. There is what is called an *umbrella standard*. This is known as EIA/TIA 568A (being revised as 568B). Within this, there are more specific standards for horizontal cables, patch cables and link testing.

There are also international standards such as ISO 9000. In Europe, there has been more concern about the health effects of large amounts of active cabling, so there has been more of a focus on shielded cabling. Level 6 and Level 7 cable are being developed. These are UL (Underwriter's Laboratories) designations for shielded cable and may be 22 gauge (fatter) than the more typical 24 gauge.

As the marketplace takes advantage of newly developing technology, the cable standards committees and manufacturers need to keep up to ensure that the cable to accommodate the speeds demanded will be available. Typically, the manufacturers respond first and the standards committees take longer to come to agreement. Cables knows as *gigaspeed* cables are now becoming available. A gigabit is one million bits.

The following information reviews a cabling standard. It was obtained from an article by William C. Spencer of Network & Communications Technology, Park Ridge, NJ (201-307-9000) in *Business Communications Review Magazine*, Hinsdale, IL (708-986-1432). The article was contributed for use in this book by Anthony G. Abbott, president of Comware Systems, Inc. (203-326-5500), a telecommunications management software company whose products include cable management systems.

In 1993, the American National Standards Institute, the Telecommunications Industries Association and the Electronic Industries Association formally approved and published their Administration Standard for the Telecommunications Infrastructure of Commercial Buildings, numbered EIA/TIA-606. This administrative standard follows and conforms to the Commercial Building Telecommunications Wiring Standard (ANSI/EIA/TIA-568, published in July 1991) and the Commercial Building Standard for Telecommunications Pathways and Spaces (ANSI/EIA/TIA-569, published in October 1990).

Although not as widely known as EIA/TIA-568 and 569, the new EIA/TIA-606 and 607 standards will have far-reaching effects on the telecommunications industry. People in charge of network management in organizations will see these standards as the foundation upon which they will build future network configurations and management systems. Those preparing Requests for Proposals will include technical specifications based on the Standard when defining documentation and identification requirements for new structured wiring systems.

System integrators, contractors, designers and installers will have to understand the 606 and 607 standards in order to respond to the new requirements which will be included in future RFPs. Software developers, who offer configuration management and telecommunication administration software systems, will have to be certain that their applications meet and comply with the new standard.

Administering a telecommunications infrastructure includes tasks such as documenting and identifying all cables, termination hardware, cross-connects, cable pathways, telecommunication closets, work areas and equipment rooms. In addition, an administrative system needs to: (a) provide reports that present telecommunications information in a useful format; (b) include drawings of the telecommunications infrastructure for design, installation and management purposes; and (c) document changes to the system with trouble tickets, service requests and work orders.

The administrative standard does, in fact, deal with *all* the components of the telecommunications infrastructure. This standard supports electronic applications such as voice, data, video, alarm, environmental control, security, fire and audio. The purpose of the 606 and 607 standards is to provide a uniform administration scheme that is independent of applications – for applications may change several times throughout the life of a building.

Three major administrative areas covered by the new standards are: pathway and space; wiring system; and grounding and bonding administration. The standard also defines specific requirements for labeling and color coding as well as including symbols recommended for use when preparing telecommunication infrastructure drawings. The concept is to establish identifiers in the form of labels that specify the content of various records and that define the linkages between records. The standard then describes exactly how to present the information needed to administer building wiring, pathways and spaces, and grounding and bonding.

Mandatory and advisory criteria are also included. Mandatory criteria required of recordkeepers specify the absolute minimum acceptable requirements and generally apply to safety, protection, performance and compatibility. Optional advisory criteria, which are considered to be above the minimum requirements, are viewed as desirable enhancements.

As part of the record assigned to each element of the telecommunications infrastructure, the standard specifies unique identifiers. Enclosed identifiers may include: additional information-cable; termination position; work area or closet location.

Labels, including these identifiers, must meet the legibility, defacement, adhesion and exposure requirements of Underwriters Laboratory 969 and should be affixed in accordance with the UL-969 standard. Bar codes, when included on labels, must use either Code 39, conforming to USS-39, or Code 128, conforming to USS-128. Labels must also be color-coded to distinguish demarcation points and campus, horizontal, and riser or backbone termination points.

Pathways must be labeled at all endpoints located in telecommunication closets, equipment rooms or entrance facilities. All horizontal and riser/backbone cables must be labeled at each end. All splice closures must be marked or labeled. Termination hardware, including termination positions, must also be labeled, except where high termination densities make such labeling impractical. The telecommunications main grounding busbar, as well as each bonding conductor and telecommunications grounding busbar, must be marked or labeled. Finally, each telecommunications space, whether telecommunications closet, equipment room or work area, must be labeled.

Each record defined in the standard must contain certain requirement information and required linkages to other specified records. Linkages define the logical connections between identifiers and records. Identifiers may then point to more than one record. Descriptions of optional information and linkages to other records outside the scope of the standard are also included but are not meant to be inclusive or complete. There is no question that properly designed administrative systems will have to incorporate many of the non-mandatory advisory elements included.

Combining both physical and logical information is important for telecommunications administration, especially when generating trouble tickets for network fault management and when generating work orders for adds, moves and changes. Being able to quickly determine which circuits are available, reserved, in use or out of use is an important part of telecommunications infrastructure management. Also, in order to associate various applications with the telecommunications infrastructure, user codes identifying and linking circuit information, such as voice or data, might be included.

The following reports are recommended by the standard: pathway, space, cable, end-to-end circuit, cross-connect and grounding/bonding summary reports. The recommended content of the reports includes:

- *Pathway reports* – list all pathways and include type, present fill and load.

- *Space reports* – list all spaces, types and locations.
- *Cable reports* – list all cables, types and termination positions.
- *End-to-end circuit reports* – trace connectivity from end-to-end and list user codes, associated termination positions and cables.
- *Cross-connect reports* – list all cross-connections within each space.
- *Grounding/bonding summary reports* – list all grounding busbars and attached backbone bonding conductors.

Additional and optional information can be presented in these reports. The reports described are not all-inclusive. Many others not mentioned in the standard would normally be included as part of a properly designed telecommunications infrastructure administration system.

Conceptual and installation drawings are considered input to the final record drawings which graphically document the telecommunications infrastructure. While the standard doesn't specify how the drawings are created, in most cases they will be prepared using a computer-aided design system, either a separate software product or a telecommunications administration system that incorporates CAD functionality.

The record drawings must show the following: the identifier as well as the location and size of pathways and spaces; the location of all cable terminations (work areas, telecommunication closets and equipment rooms); and all backbone cables. Drawings which show the routing of all horizontal cables are desirable. The standard includes symbols that may be used when preparing these drawings. Ideally, record information should be accessible when one is viewing the record drawings.

It is mandated that all wiring, termination and splice work orders be maintained for telecommunication repairs, adds, moves and changes. The work-order document must include cable identifiers and types, termination identifiers and types, and splice identifiers and types. The work-order process should be used to update the administrative records. **In day-to-day telecommunications administration, this is the most important requirement set forth in the standard.** If the system records are not immediately updated when a work order is completed, the administrative system will quickly become outdated and useless.

The International Organization for Standardization's Network Management Forum identifies five functional network management areas: configuration management, fault, security, performance and accounting management. Configuration management is the core of the four other network management areas and comprises the following management elements: in-use and spare-part equipment inventory management; cabling and wiring management; circuit management; tracking, authorizing and scheduling adds, moves and changes; trouble ticketing network faults; user and vendor management; and documenting current network configurations.

If the mandatory and advisory criteria are included, the ANSI/TIA/EIA-606 and 607 Administration Standards for the Telecommunications Infrastructure of Commercial Buildings cover most of these configuration management elements. Since the infrastructure can be thought of as the collection of those components that provide basic support for the distribution of all information within a building or campus, the telecommunications administration standard must now be viewed as the basis upon which all future network configuration management systems will be built.

Implementing a telecommunication administration system requires a great deal of thought and planning. There are many important reasons why organizations should implement a physical layer configuration, design and telecommunications administration system:

— To determine what cables, conductors or fibers, and circuits (PBX, Ethernet, Token Ring, etc.) are free, in use, and out of use, and what circuits and users are assigned to them.

— To maintain a documentation and identification system for the implementation of an equipment and cable disaster recovery plan in case of fire, explosion, flood, or other natural or man made emergency.

— To identify what equipment is in use, spare and out of use, and to document and maintain equipment connectivity.

— To update and manage network faults, adds, moves and changes, and to maintain work order records for all equipment, users, circuits and cable paths.

— To reduce the amount of network or LAN downtime.

— To decrease labor costs by eliminating the need to trace undocumented circuits each time an add, move, change or network fault occurs.

— To increase confidence in the structured wiring systems that organizations use to downsize applications from mainframe platforms to client/server environments.

— To generate management reports and perform detailed network analysis on all equipment and cabling systems.

— To transfer users from one network to another as part of the effort to manage overall network performance.

— To document and maintain cable and circuit test data.

— To manage important vendor relationship: purchasing, technical support, returns and service.

— To administer names and addresses and track all network equipment, cables and circuits for departments, users, managers and technicians.

— To design new networks within the infrastructure and to produce reports and analyses detailing equipment and cabling requirements.

There are two major challenges to implementing the 606 and 607 standards. The first is finding people who have the training and motivation to understand and implement them. The second is the substantial cost of labor needed to achieve the level of detail required. Historically, upper management of most organizations has been unwilling to invest in labeling and recordkeeping as a means of avoiding future disruptions in service and future costs. This is particularly true if the infrastructure is already in place and the records must be assembled (as opposed to a brand new installation). When customers begin to demand better recordkeeping on the part of their telecommunications suppliers, things may change.

The focus of this chapter has been cabling within the business premise, but cabling also carries most transmissions between locations. The next chapter talks about outside lines and transmission.

📖 For more information on Cable see Cable Inside & Out by Frank W. Horn and Understanding Fiber Optics by Jeff Hecht (order from 1-800-LIBRARY or www.telecombooks.com).

CHAPTER 9

Outside Lines and Transmission

The terms *outside lines, circuits* and *network services* all refer to the same thing. These are the different types of lines you may install from the local or long distance telephone companies to connect your organization to the outside world or your other sites.

Circuits To Connect Your Telephone System To The Local And Long Distance Network

A *circuit* is the physical connection between two communications devices. Some circuits are permanent connections between two points and may be known as *fixed, dedicated* or *point-to-point circuits*. These are also known as *leased lines*. Other circuits are temporary and may be called *switched circuits* since the temporary physical connection between two points is accomplished by a switching system.

In order to connect your telephone system to the rest of the world, it is necessary to order outside lines, also known as *trunks*. Here are the types of outside lines you may connect to your telephone system.

Combination Trunks (Both Way Trunks)

A trunk is *one* outside telephone line. Since a trunk sounds like something big, it is often thought to represent many lines, but it does not. The term *trunk* refers to an outside central office dial tone line that comes into a PBX. A *combination trunk* or *both-way trunk* indicates that the line can be used for both incoming and outgoing calls. Callers that dial into the main telephone number of a company and are answered by the switchboard may be calling on one of a group of combination trunks. This is set up in what is called a *hunt group* or *rollover sequence*, also called *ISG* or *Incoming Service Grouping.* This means that when the first line is busy, the call comes in on the second line, and then on to the third if the second is busy, and so on until the end of the hunt group. Employees inside the company may place outgoing calls on that same group of combination trunks when they *dial 9.* Using 9 is an historical convention, but it is technically possible to use any number or several numbers as an *access code* to dial out. Although combination trunks permit both way calling, most telephone systems have the capability to be programmed so that combination trunks are used for only incoming or only outgoing calls. You may hear the expression *DOD* or *direct outward dial trunk.* This is typically a combination trunk used for outgoing calls only.

Combination trunks are usually *ground start trunks.* In order for the connection to the telephone company central office to be completed, the switch in the central office must receive a signal from the PBX on your premises. The signal results from a momentary grounding of the circuit. When the signal is received, the connection is completed and you hear an outside dial tone. The ground start trunks provide what is known as *supervision.* When the trunk is seized by someone dialing 9, the PBX looks ahead to see if the same trunk may have also been seized by the central office to send a call to the PBX. If so, the PBX will know not to use that trunk.

It is possible to use *loop start trunks* in a PBX. Loop start trunks are more likely to be used on a key system. When you press down the line button, the dial tone is there. You do not need to dial 9 and the telephone system is not signaling to the central office. If you use loop start trunks on a PBX, there is a risk that someone calling in on a trunk will collide with someone who has just dialed 9, selecting the same trunk to place an outgoing call. This type of collision is called *glare*. This will not happen on a ground start trunk.

Most groups of combination trunks have a separate seven digit telephone number associated with each trunk. If the central office runs out of telephone numbers, it may provide *coded trunks*. These have no associated telephone number and are reached only by dialing the main number and then rolling over as part of a hunt group if the main number is busy.

A combination trunk is a switched circuit. It is one type of dial tone line (see *dial tone lines* later in the chapter).

Direct Inward Dial (DID) Trunks

Direct inward dial trunks are used for incoming calls only. Although in some parts of the United States (like New York City) they cost up to three times as much as combination trunks, they do provide a special function. Using direct inward dial trunks enables everyone within a company to have his own separate seven-digit telephone number. The first three digits of everyone's number are the same. The telephone numbers that work with DID trunks are purchased from a local telephone company along with the trunks in blocks, usually of 20 or 100. To the caller, these are indistinguishable from regular telephone numbers. Large companies using DID may assign the first seven digit DID number of the group as their main telephone number that rings into the switchboard.

The point of DID is that you do not need a separate trunk for each telephone number. You may have 10 DID trunks with 100 DID directly dialable telephone numbers. The assumption is that no more than 10 telephone numbers will be called at any given time. If an eleventh call comes in, the caller hears a busy signal.

When a DID telephone number is dialed, the central office recognizes the number and sends it to the PBX. Once it reaches the PBX, the last 3 or 4 digits (important to specify which when ordering) are repeated to the PBX, which then knows to which telephone within the office the call is to be directed.

DID trunks in the U.S. are *wink start trunks*. The wink is a signal that the PBX sends to the central office to let it know that the PBX is ready to receive the call and the digits dialed. In Europe, DID trunks are *immediate start* meaning there is no wink.

Once a call has reached the DID extension, the PBX treats it just like any other extension. When selecting a DID extension on the telephone instrument and dialing 9, you are not dialing out on a DID trunk. You are dialing out on one of the combination trunks with which your system is also equipped. DID trunks are for incoming calls only.

A DID trunk is a switched circuit. If DID trunks are delivered on a PRI ISDN circuit (discussed later in the chapter) they may also be used for outgoing calls. This is sometimes called a *two-way DID*.

Tie Lines

A *tie line* is a point-to-point line between two telephone systems (usually PBXs). You may hear the term *dial repeating tie line* or *E&M* tie line (stands for Earth & Magneto or Ear & Mouth depending on whom you ask), which is the same thing as dial repeating.

The idea behind a tie line is that you can dial the extension of someone on a PBX at a distant location without placing an outside call through the public network and without talking to the switchboard attendant at the distant PBX. A tie line may be accessed by dialing one or two digits into your PBX; for example, dial 8. Once you dial 8, you hear dial tone (also called *drawing dial tone*) from the distant PBX. You may then dial any extension within that PBX. If the distant system is programmed to allow it, you may also be able to dial 9, instead of dialing an extension number. This enables you to place an outgoing call through the distant PBX. For example, if the tie line connects Seattle to Los Angeles, the person in Seattle may dial through the Los Angeles PBX over the tie line and be charged only for a local call in L.A.

Tie lines usually connect separate offices of the same organization. A single tie line can handle only one conversation at a time. If there will be many calls going back and forth between the two PBXs, you need a sufficient number of tie lines to handle the volume. You can put the tie lines on a high capacity circuit such as a T-1. If you have a network connecting more than two PBXs, you may call over successive tie lines which route through one or more PBXs before reaching the final destination.

Many PBXs are intelligent enough to recognize the location of the three- or four- digit extension number you have dialed. They send the call automatically to the correct PBX and telephone at that site. In this case, there is no need for you to dial an access code. The PBXs are still sending access codes to each other to route the call, but this is transparent to you. This capability to dial among PBXs on a network by simply dialing the extension number is sometimes called a *uniform numbering plan.*

A tie line is sometimes called a *tie trunk.* It is a dedicated circuit between two points.

Off-Premise Extensions

An off-premise extension is another type of point-to-point line, usually with a PBX on one end and a single line telephone or a key system on the other. The idea is that if a call comes into the PBX, that call can then be extended to a distant location, across the street or across the country. To the caller it appears that the person is at the location he called. The person at the off-premise extension can pick up the telephone and dial any extension within the distant PBX, just as if he was on site. He can also, if set up to do so, draw dial tone and place an outside call through the PBX. If the off-premise extension is on a key system rather than on a single line telephone, it will appear on a separate button on one or more of the telephones in the key system. There may be one or many off-premise extensions to the same or different locations.

Local telephone companies have regulations concerning off-premise extensions. For example, if you want to have an off- premise extension from a business and have it terminate at a residence, you must already be paying for a separate residence telephone there.

Off-premise extension telephones are typically analog. Digital telephone systems have a distance limitation in terms of how far a proprietary digital telephone can be located from the PBX control cabinet. Some system manufacturers are designing more distributed systems. These control components of the PBX, such as a shelf or a circuit board and may be at a distant location to provide control for the telephones there.

With newer telephone systems running on the company LAN (local area network) or WAN (wide area network connecting multiple LANS), if a dedicated network is already in place the telephones may be plugged into it and “piggyback” on circuits already in place. No additional control equipment from the telephone system manufacturer is needed at a remote site, even if there is just a single telephone.

An off-premise extension is a dedicated circuit.

Foreign Exchange Lines

A foreign exchange line is a type of dial tone line that is brought into a PBX from a distant central office or exchange (the exchange is the first three digits of the telephone number). These three digits, within each area code, are associated with a particular central office and therefore a particular geographic area.

Foreign exchange lines are used for two different purposes. If a company has a heavy concentration of telephone calls to a particular city, it may save money to install one or more foreign exchange lines from that city into their PBX. When someone calls that city, the PBX selects the foreign exchange line and the call is billed as a local call since the telephone line actually comes from the central office in that city.

Companies in cities with many different area codes within a small geographic area, such as Los Angeles, may particularly benefit from the use of foreign exchange lines. As the cost per minute for making long distance calls has been reduced, so has the need for using foreign exchange lines in this manner.

The more common use of a foreign exchange line is to give a company a local presence in another city. For example, a company is listed in the Chicago telephone directory and Chicagoans place a local call to reach it. The callers are not aware that the calls are being answered by a PBX located in Milwaukee through a group of foreign exchange lines.

Foreign exchange lines are becoming less common as the cost of toll free incoming (800,888) lines drops. More organizations are opting to give callers toll free numbers to call and seem less concerned with the local presence offered by the foreign exchange lines.

Telephone company exchanges used to be easily identifiable as being associated with a certain city or neighborhood, but that is no longer the case as many exchanges may now be in use in the same location.

A foreign exchange line is a switched circuit.

T-1 Circuit

The above circuits can be delivered to your premises separately with each trunk brought in on a single pair of copper wires. Any of these may also be delivered on what is called a T-1 circuit. The T-1 has the capability for 24 separate *channels*. Each channel of the T1 may carry a certain type of trunk such as a combination trunk or a Direct Inward Dial. A T1 can also be used to carry point-to-point type of dedicated lines. The entire T1 is usually a *4 wire circuit* and is sometimes brought into the premises on a device known as a *smart jack*.

In order to work, the T-1 needs a *multiplexer* at either end to break the 24 separate conversations down and reassemble them at the other end. Sometimes the multiplexer is external to the telephone system. One common type of external multiplexer is called a *channel bank*. The channel bank may be purchased or can be rented from the carrier providing the circuit. Consider each option from both cost and service perspectives. Who will you call first if the circuit goes down? Most PBXs also have the capability for a *T-1 circuit board* which is a multiplexer inside the PBX.

You may have different types of trunks combined on a T-1. For example, you can have 12 combination trunks and 12 DID trunks.

Digital services have evolved based upon the digital signal known as *DS0* (digital signal zero) which is 64 Kbps. This is the bandwidth for a voice grade circuit. *DS1* (associated with a T-1) is 1.544 Mbps, which is twenty-four 64 Kbps channels. There is also a *T-3*, which is 44.73 Mbps (the equivalent of 28 T-1s).

On a T-1, signals are multiplexed so that two pairs of wire can carry 24 separate voice or many (depending upon the speed required) data conversations. These same two pairs of wires would carry only two voice conversations using analog technology. With each conversation using 64 Kbps and another 8 Kbps being used for control signals, the total T-1 capacity is 1.544 Mbps (million bits per second).

You may also hear the term *fractional T-1* which means that you are getting a circuit with some fraction or increment of the total T-1 capacity. For example, the circuit may provide only twelve 64 Kbps channels. Actually, the entire T-1 is in place, but some of the channels in the multiplexers are not activated.

T-1 Circuit Applications for the PBX

T-1 TO THE LOCAL TELEPHONE COMPANY CENTRAL OFFICE

As the price of a T1 circuit continues to decrease, many organizations are choosing to bring in their outside lines to the local telephone company over the T-1. It does not cost any more and can often cost less than bringing 24 outside lines in over 24 separate pairs of wires.

In this instance, the other end of the T-1 is at the central office of the local telephone company. They are responsible for the multiplexing capability at their end.

T-1 TO CONNECT TWO OFFICES OF THE SAME ORGANIZATION

Another common use of a T-1 is to connect two locations of the same company within a network. This may be used for 24 tie lines connecting 2 PBXs or it may be used for a combination of voice and data circuits. In the latter case, it is advisable to terminate the T-1 in a multiplexer outside the PBX. It is likely that the people using it for data will need to monitor the line and may want more sophisticated diagnostic capabilities than are available when the T-1 terminates in a circuit board in the PBX. Extricating some of the T-1 channels from the PBX to be used for data can be cumbersome.

T-1 TO YOUR LONG DISTANCE CARRIER

Still another T-1 use is to connect a company to the nearest switch or *POP (point of presence)* of the long distance carrier. The long distance companies encourage this since it tends to secure the customer, making it more difficult to change to another carrier. Used in this way, the T-1 may be called an *access line.* The T-1 connection provides 24 two-way paths or channels on which calls can be placed through the long distance carrier network. Companies with T-1 connections may pay a lower cost per minute for their telephone calls.

If you buy your local telephone service from the same company that sells you long distance, both types of services use the same T1. This is more efficient than having separate circuits for local and long distance calls.

📖 For more information on T-1 see The Guide to T-1 Networking by William Flanagan (order from 1-800-LIBRARY or www.telecombooks.com).

ISDN LINES

ISDN is described in the next section since it is also used outside of the PBX. Most PBXs have the capability to accept an ISDN PRI circuit and require a separate circuit board to do so. Few accept the BRI type of circuit.

Telecommunications Circuits Not Typically Connected To The PBX

Dial Tone Lines

This term generally refers to any outside line which, when accessed, gives a *dial tone* from the central office switch of the local telephone company. The central office is the point at which all telephone numbers are created.

You may also hear the term *POTS line* (plain old telephone service), *auxiliary*, *aux line* or *private line* used to describe a dial tone line. It is delivered to your telephone equipment room on two copper wires and may be terminated either on a separate modular jack or on the demarc in the telephone equipment room. PBX combination trunks are also a type of dial tone line, but not usually referred to as such.

A dial tone line can be used for receiving or placing calls. It has a telephone number associated with it, also called a *number assignment* or *line assignment*, provided by the local telephone company. This is a seven digit number preceded by the three-digit area code. The first three digits of the seven numbers are called the *exchange*, the *central office*, or the *NNX*. In the case of a dial tone line, the telephone number assignment is also the circuit number. Other types of circuits have different types of circuit numbers usually including letters and not in the seven-digit format. For example, 96OSNA111222 is a circuit number for a point-to-point circuit.

A dial tone line may be a *loop* start or *ground start line.* When ordering an outside telephone line, you will be asked if the line is to be ground start or *loop start.* Dial tone lines coming into most key telephone systems or directly to a telephone, fax or modem, are loop start. This means that when you access the line, the physical loop connecting you to the local telephone company central office becomes a complete circuit for sending and receiving calls. (*Tip and ring* are names for the two wires that complete the talk path.) You hear the dial tone immediately and can place your call. Most PBXs require ground start lines, in which the central office switch is looking for a signal from the PBX before the circuit is completed and dial tone is delivered.

When you are at a PBX telephone, you may lift the receiver and hear a dial tone. This is not the dial tone being sent from the local telephone company, but rather the dial tone sent from the PBX on premises. To place a call to the outside world, you typically dial 9 and hear a second dial tone that will be the one from the local telephone company. You may then dial the outside telephone number you are calling. As mentioned above, a ground start PBX trunk is usually not called a dial tone line, even though it is one.

Point-To-Point (Dedicated) Voice Communications Lines

MANUAL AND AUTOMATIC RINGDOWN CIRCUITS

These are circuits that permanently connect two points, when instantaneous communication is important, such as in the case of two brokerage traders needing to speak immediately to execute a trade. The person at one end presses a button on his telephone associated with the specific line; another button at the distant end flashes or rings, indicating that the line is to be answered. If the person originating the call needs only to depress the button associated with the line to signal the other end, then the signaling is automatic and the line is called an *automatic ringdown line.* If the person must depress the line button and then press a separate signal button to signal the other end, the line is known as a *manual ringdown line.* When a point-to-point line is rented from the local telephone company, the connecting point between the two locations will still be at the local central office rather than running directly between the two end points. The cost of these lines is based upon mileage, which considers the distance from each end point to the central office. Different parts of the U.S. have different methods for billing point-to-point lines.

There may also be long distance, sometimes called *long haul,* ringdown private lines. Many brokerage firms on Wall Street in New York have lines going to Chicago and the West coast. In this case, the lines are rented from one of the long distance carriers, although the local part of the circuit may still be delivered to the premises on cable from the local telephone company.

Centrex

Centrex is a type of telephone service offered by the local telephone company in most areas of the U.S. It goes by different marketing names in different areas including Centrum, Essex and Intellipath, to name a few.

The idea behind Centrex is that your organization does not need to buy a PBX.

The switching of your calls takes place at the telephone company central office instead, with the extensions coming out to your premises through cables under the street or overhead. You use system features such as call transfer and call conference, which are capabilities of the central office switch.

It is possible to use almost any telephone system with Centrex service, but some make more sense than others. The best solution is usually to purchase the telephones that are made to work with the central office switch. Northern Telecom makes a central office switch. If your Centrex service is delivered from this type of system, you may want to buy Northern Telecom telephones to realize the maximum capability for using the system features. Lucent Technologies also makes central office switches, as do Siemens, NEC, Ericsson and others.

Centrex is a particularly good service for providing telephone service to multiple locations of the same company or organization within the geographic area served by a single central office. For example, a municipality or a university with many separate buildings may benefit from Centrex service. In cities, the Centrex serves a smaller geographic area than in areas less densely populated.

With Centrex service, each telephone in the system has a separate telephone number and may carry extra monthly charges for certain system features. Over time, it tends to be a more costly system than an on-site PBX, but some organizations like the idea of not having to manage their own PBX.

If you decide to use Centrex, you may not have as much control as you would have over a PBX on your premises. Also, you may not always have the very latest in system capabilities. Before the local telephone company invests in upgrading the central office switch, there needs to be sufficient demand to justify the expense of the upgrade. Since some local telephone companies are regulated, they must also apply to the public utilities commission in their state for permission to offer or *tariff* a new service.

"The Network"

Any group of circuits and their associated hardware is called a *network.* There are many different types of networks. A single site with a variety of lines connecting it to the outside world may be referred to as a network. More commonly, network is used to describe multiple sites linked together via telecommunications circuits. The network can be for data only or voice only. Many networks now combine voice and data along with Internet access.

ISDN

ISDN stands for Integrated Services Digital Network. This is a collection of standards and protocols for digital communications. The benefit of ISDN service is that it provides more advanced capability over the same pair of wires used to deliver a regular dial tone line to your home or business.

In TELECONNECT magazine, Madeline Bodin does a good job of explaining the two things that ISDN lines enable you to do which your POTS lines (plain old telephone service) do not.

First, ISDN lets you transmit more information from one point to another. It gives you more *bandwidth.* Comparing the information being sent on the telephone line to water, a regular telephone line would be like a straw and an ISDN line would be like a 4 inch diameter plumbing pipe.

Second, ISDN lets you control the information going across the line from outside the pipe, called *out of band* signaling. By doing this, more information about each call is delivered, the information is more secure and a number of limitations are removed. (Note: Currently, long distance companies accomplish this *out of band signaling* over a separate dedicated network called *Signaling System 7* in order to open and close switches and route calls.)

The local telephone companies use *in band* signaling right on the same line as the transmission with POTS lines. Signals such as tones take up transmission time that could be more efficiently used for transmitting voice and data.

There are two ISDN interfaces: *Basic Rate Interface* and *Primary Rate Interface.* Basic Rate Interface (*BRI*) (Figure 9.1) is usually described as *2B+D.* That means it has two *bearer channels* and one data channel or D channel. The B channels carry the content of the call that may be voice, high speed data or video. Each B channel carries 64 kilobits per second. The D channel carries the call delivery information, such as the telephone number of the calling party. The D channel carries 16 Kbps.

Figure 9.1

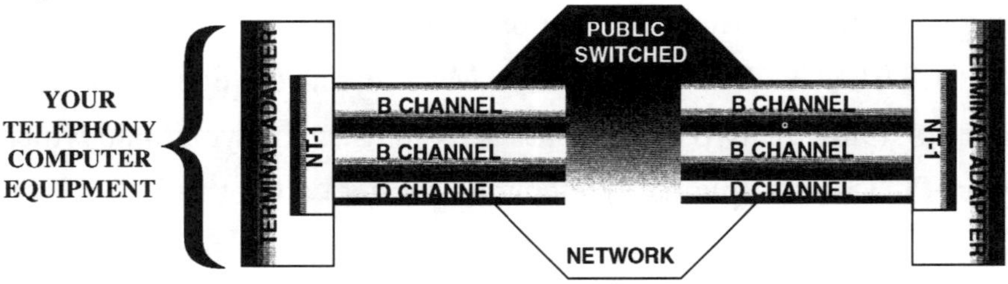

Compliments of TELECONNECT MAGAZINE

The Primary Rate Interface (*PRI*) is described as 23B+D for a total of 24 channels, the same as a T-1. PRI is what happens when ISDN meets T-1.

In addition to the ISDN service and compatible telephone and computer equipment you also need a *Terminal Adapter* and a *Network Termination Device* known as *NT-1.* Sometimes the NT-1 is built into the terminal adapter.

Applications for ISDN include:

- Ability to send and receive data rapidly-makes it popular for credit card processing.
- Internet users do not have to wait a long time for graphics to appear on the computer screen.
- *ANI (Automatic Number Identification)* delivered on ISDN. Enables user to capture the telephone number of the person calling, enabling you to accept, reject, ignore or redirect incoming calls. *Note*: You do not need ISDN to get ANI, but ISDN delivers it sooner.
- Two users can be talking and exchanging computer data at the same time, making ISDN valuable for Help Desks.
- Desktop video conferencing is accomplished by bonding together the 2 B channels to provide sufficient bandwidth for video transmission.

📖 For more information on ISDN see ISDN: A Practical, Simple, Easy-to-Use Guide to Getting Up and Running on ISDN by William Flanagan (order from 1-800-LIBRARY or www.telecombooks.com).

DDS Circuits

DDS Service (Dataphone Fixed Digital Service) is a basic digital circuit rented from AT&T (their terminology) or local telephone companies. The same type of service is available from other long distance carriers. One source of confusion in the telecommunications industry is that different carriers (long distance and local) give different names for to the same types of services. DDS can be subscribed to only as one fixed speed at a time up to 56 Kbps.

DDS circuits (typically in pairs: one send and one receive) usually call for two separate dial tone voice-grade lines with modems known as *dial back-up lines* that are used if the DDS lines fail.

As the price of T-1s drops, this type of service is beginning to disappear.

Switched 56

Another type of digital service is called *Switched 56.* This is for voice and data applications that require high transmission rates but do not require a connection between the two points at all times. The connection is established through dialing only when it is needed.

Switched 56 can also be used as a backup for other digital lines.

Asynchronous Transfer Mode (ATM)

ATM (Asynchronous Transfer Mode) appears to hold great promise for transmitting very high volumes of information over a circuit at very high speeds. One ATM circuit can handle 155 megabits per second. This is a digital service and uses fiber optic cable. It transmits the data in cells incorporating the latest high-speed assembly technology. You may also hear the term. *SONET*, which stands for *Synchronous Optical Network*, which is necessary for proper and reliable transmission and reception of ATM. SONET is an optical interface standard that enables transmission products from different vendors to network with each other.

📖 For more information on ATM see ATM: Asynchronous Transfer Mode: The Complete Guide by William Flanagan (order from 1-800-LIBRARY or www.telecombooks.com)

Transmission

Transmission is a complex topic. Here we introduce some basic concepts.

Transmission has to do with getting information from one place to another, whether it's across the office or thousands of miles away. The connection between the two points is referred to as the *circuit.* One of the challenges is to make distant locations seem as if they are just across the office. The information being transmitted may be a voice, written information, still pictures, moving pictures or some combination of these.

Here are some of the things to think about, relating to telecommunications transmission.

Transmission Medium

The communications signal is a low-voltage electrical signal that must have some means of getting from one point to another.

This is the *transmission medium.* The medium may be a physical connection such as a cable. As we pointed out in Chapter 8, there are different types of cable made out of different materials, but most commonly copper (twisted pair and coax) or glass (fiber optic cable). There is also plastic fiber optic cable called *PFO*, polymer fiber optic. A cable can be used to connect devices several feet from each other. Cable is also run on the ocean floor to connect the continents.

The second most commonly used medium is space. Communications signals sent via microwave or satellite are transmitted through the air or through space outside the atmosphere.

Water is also a communications medium, such as with submarine to submarine communications, but it is not used to carry your average business communications signals.

In general, telecommunications signals travel either through a cable or through space. The ones that go through space have often started out in a cable and may end up in another one before they reach their final destination. As wireless communications becomes more widespread, the need for cable is diminishing, but it is not likely that it will go away.

All of this cable running through the office, under the street and around the Earth is useless as a transmission medium for communication signals unless there is some hardware at each end of the circuit and some more in the middle.

Personal Computers, Network Computers and Dumb Terminals

At each end of the circuit there must be some type of communications device to send a signal and another one at the other end to receive it. This may be a telephone, a fax machine, a terminal or computer (PC) with a modem, a multiplexer or a video transmission device.

It is becoming increasingly common to have a personal computer as the device at the end of the communications circuit. Within the next decade the computer may replace the telephone as the instrument for handling voice communications (see Chapter 12 on Computer Telephony).

Terminals with no built in intelligence, known as *dumb terminals,* used to be the most prevalent type. They were connected to a central *mainframe* computer where both the intelligence and information resided. The characters traveled to the mainframe individually as they were typed in by the user.

This type of transmission is called *asynchronous transmission*. The signals are sent over the line in a random fashion with added signals called *start bits* and *stop bits*. These start and stop bits tell the device at the distant end when to start and stop sampling the data coming across the circuit. The information is transmitted in digital form (different combinations of ones and zeroes representing all letters, numbers and symbols).

There is also *synchronous transmission.* In this case, digital information is sent between two devices that are in a specific time relationship, synchronized by a master clock. This is typically used between mainframe computers. It enables blocks of information to be sent, rather than individual characters one at a time.

While some dumb terminals are still in use, most have been replaced by PCs (personal computers). PCs communicating with each other or with mainframes use an asynchronous type of transmission.

The terms synchronous and asynchronous transmissions are becoming dated. Instead, we refer to a particular technology such as ATM or protocol *TCP/IP* (*Transmission Control Protocol/ Internet Protocol*) or *thin computers.*

Computers called *NCs (Network Computers)* may replace some PCs. They cost less since they are using the intelligence and information residing at some other location in the network.

Modems

Modem stands for MODulator and DEModulator. One type of modem sits at each end of an analog telephone line that converts digital signals to high and low audible tones. These tones represent zeroes and ones, enabling the digital format to be carried on an analog line.

Since most telephone lines in place are designed to accept analog signals, the modems enable digital signals to be transmitted over an analog line. The tones are discrete. If not kept discrete, a series of ones would be indistinguishable.

This is achieved using varying techniques. The modem demodulates the stream of tones at the far end of the circuit, converting the ones and zeroes into a format that can be understood by the device with which you are communicating. There are many different modulation schemes.

You may have seen one of the earliest types of modems developed called an *acoustic coupler*. The acoustic coupler is a device with two built-in rubber cups into which a standard telephone handset (receiver) fits. After dialing the telephone number of the distant end device, the communication link is completed by placing the handset into the coupler to form an acoustic connection, rather than an electrical one, between the modem and the telephone line. This device was developed in the pre-Carterfone Decision days when the Bell System was concerned about any foreign devices being hardwired onto the telephone lines. Acoustic couplers are asynchronous modems.

The basic components of a modem are a transmitter, a receiver and a power supply. The transmitter includes circuitry for, among other things, *modulation*.

The telephone line carries a signal in the form of a sine wave. The three parts of the sine wave which can be manipulated or modulated by the modem to represent data are frequency, amplitude and phase. Modulation techniques have been developed around each of these components. The higher the speed of the modem, the more complex the scheme needed to impress information on the sine wave. The more complex circuitry leads to greater cost.

Here are the modulation techniques:

— *Frequency Modulation* is used for low-speed asynchronous transmission. The number of waves per unit of space on the circuit is varied while the height or amplitude of the waves is kept constant. The one bit is represented by two waves and the zero bit is represented by four waves. (Remember, the computer can only understand ones and zeroes. The *bit* is the one or the zero.)

— *Amplitude Modulation* varies the height or amplitude of the waves while keeping the frequency constant. The waves representing the one bit are taller than the waves representing the zero bit. This technique is often used to transmit data between 300 bps and 1200 bps. Amplitude is seldom used in telecommunications. AM radio stations use amplitude modulation.

— In *Phase Modulation*, the normal sine wave is used to represent the one bit and a sine wave that is 180 degrees out of phase, the mirror image of the original wave, is used to represent the zero bit. Phase modulation is most common in high speed modems.

There are more complex modulation schemes that combine aspects of the above three and may include error-correcting capabilities.

The following lists some features and options which can provide modem flexibility. As with most telecommunications equipment, some manufacturers' optional features may be standard for others.

— *Multi-port capability* means that more than one outside telephone line can be handled by the modem at the same time.

— *Multiple speed selection* permits you to continue transmitting even while the quality of the telephone line is degrading, by falling back to a lower transmission speed, thus reducing the error rate.

— *Dial back-up* allows you to switch from your dedicated leased line to a dial line or lines in the event that the leased line fails.

— *Voice and data capability* allows you to alternate voice transmission with data transmission on the same line.

— *Self-testing diagnostics* perform local and remote testing to determine if there is a problem with either the line or the modem.

— *Autocall* automatically places calls for a modem on the dial network.

— *Auto-answering* automatically answers incoming calls to the modem.

— An *adaptive rate system* built into the modem allows it to continuously sense varying line conditions and adjust according to the highest possible speed.

— *Echo suppression* is an advanced technique that minimizes echoes and the resulting distortions for 2-wire dial line transmissions over short and long distances.

— *Phase roll compensation* is a feature that automatically adjusts for frequency differences between long distance carriers.

— A *modem substitution switch* is an external option. It allows you to reroute your data through a spare modem. The spare is standing by, powered up, in the event that the other modem fails.

— An *asynchronous-to-synchronous* converter permits an asynchronous modem to operate with a synchronous modem.

— *Modem strapping options* enable the modem to switch from half to full duplex operation. *Full duplex* means transmission can be taking place in both directions simultaneously. *Half duplex* means you can either send or receive, but not do both at the same time.

The three basic criteria for selecting a modem are: the volume of transmitted data; the speed that the data is to be sent; and the distance the data needs to travel.

Volume is a function of both the characters per transaction and the number of transactions per day required to support an application. The volume of information determines not only the speed of transmission, but whether you should use leased lines or dial up lines.

Other questions to ask when you are selecting a modem include:

- Will the modem be compatible with your devices at each end?
- Will new modems be able to communicate with existing ones?
- Will the modem be used on a leased line (a fixed line connecting two ports) or a dial line (a line with a dial tone requiring you to dial a telephone number to reach the distant end)?
- Will the modem be used in a point-to-point or multi-point configuration?
- What kinds of diagnostic capabilities are needed?
- What is the acceptable error rate?
- Do you need multiple speed capability?
- How are the options added?
- Does the modem have network control capability to keep the network running despite problems?

Multiplexers

If you want multiple devices to be transmitting simultaneously, or if you want to mix synchronous and asynchronous transmissions, consider installing multiplexers in front of the modems. The *multiplexer* enables two or more signals to be transmitted simultaneously on the same circuit.

Multiplexed transmissions are so fast that they appear to be simultaneous. However, the data being transmitted is actually moving across the circuit sequentially. The multiplexer breaks each transmission into smaller pieces, sending pieces from each transmission and reconstructing them again at the other end.

A multiplexer, also known as a *mux*, is located at each end of the circuit. At the receiving end it is sometimes called the *demux (demultiplexer)*. The multiplexer is used to maximize the use of a high cost leased line by enabling it to be used for multiple simultaneous transmissions.

There are three types of multiplexing:

— *Frequency division multiplexing* is an analog technique, dividing the bandwidth into smaller channels. This is used today for cable TV signals in what is known as *broadband*. The FM dial on a radio is an example of frequency division multiplexing. This can take place in the air or on a coaxial cable.

— *Time division multiplexing* can operate in a digital (called *pulse code modulation*) or analog (called *pulse amplitude modulation*) manner. All PBXs now use pulse code modulation also called *PCM*. Older PBXs used pulse amplitude modulation.

Each device on the network is assigned a particular time slot on the high speed line whether it has something to transmit or not. The order in which the devices transmit always stays the same. The digital time division multiplexer operates by scanning the devices, sampling the first bit from each transmission and then repeating the process sampling the second bit. It continually transmits these interleaved bits across the telecommunications circuit. A limitation of time division multiplexing is that it cannot retransmit an error, because transmission is continuous. This is significant for data transmission, but insignificant for voice transmission.

— *Statistical multiplexing* is an improved form of time division multiplexing in that it maximizes the use of the lines on your network, but it can sometimes cause blocking if the statistics do not hold. The statistical multiplexers take advantage of the idle time of communications devices by dynamically allocating time slots only as the devices require them. Thus, more devices can be served on the same line.

These *stat muxes* also have buffers so that, when statistics are wrong and more devices are transmitting than statistically anticipated, the buffer will hold the information until the circuit is free to transmit it.

Hardware for Digital Circuits

Lines for digital transmission require a device called a *DSU (Data Service Unit)* or *CSU (Channel Service Unit)*. Often the DSU and CSU are built into a single unit. You may even buy an analog modem that can become a CSU/DSU with a software upgrade, should you change from analog to digital service. The *CSU/DSU* maintains a steady current on the line to control the quality of the signal being sent.

Other Transmission Hardware

Other types of hardware come into play for transmitting telecommunications signals. For devices that are nearby, the signal may travel over the medium without any help. For relatively longer distances the signal begins to attenuate and must be boosted or amplified by what is called a *repeater*.

All communications circuits also have devices called *transformers*, as do other electrical circuits. The transformer provides a variety of functions, including: stepping up and stepping down voltage; managing of line impedance; and electrical isolation via magnetic coupling, an important safety item.

Other types of hardware that send communications signals through the air are the *microwave dish* and *satellite* with its *transponders*. These satellites are in geo-stationary orbits appearing to hover at a fixed point 22,000 miles over the Equator. In reality they are orbiting at a speed in excess of 8,000 mph in sync with the Earth's rotation.

A microwave dish sends a signal to another dish or bounces the signal off a satellite. The signal is then received by another dish. A microwave dish may be several feet in diameter or the size of a three-story building. (Do not stand in front of, or even near a dish, if it is transmitting. It may reorganize some of your cells. The receiving dishes are harmless.)

We have the transmission medium and the hardware to make the medium useful. What else relates to transmission?

Two things to consider when sending communications signals is how much information can be sent at the same time, and how fast can it get there. Bandwidth refers to the capacity of a circuit for carrying communications signals. *Bandwidth* has come to mean not width, but rather the speed at which information travels over a circuit. For example, you may hear someone say that the bandwidth of a circuit is 56 Kbps (kilobits per second).

The origin of the term bandwidth was for a coaxial cable on which different transmissions are carried along different frequencies in varying distances from the center or *carrier frequency*. A coaxial cable is considered to be *broadband*.

True bandwidth refers to the range or speed of frequencies that can be carried by a circuit. The hardware at each end and in the middle determines the number of separate frequencies which can be transmitted simultaneously on one circuit.

The capacity of a particular circuit and the speed with which communications signals can travel over that circuit are a function of both the bandwidth of the circuit and the capability of the hardware used at both ends and in the middle.

If a circuit has a higher speed (higher bandwidth,) the signals do not actually move faster. It's just that more information can travel over the circuit almost simultaneously.

Another basic concept relating to circuits is whether they are simplex, half duplex or full duplex. On a *simplex* circuit, such as a radio broadcast, information travels in one direction only. On a *half duplex* circuit, such as a CB radio, the information travels in both directions, but only one way at a time. You can either talk or listen. A *full duplex* circuit, for example a telephone conversation, enables simultaneous two way transmission – you may talk and listen at the same time.

Protocols are the rules, procedures or conventions relating to the format and timing of communications transmissions between two devices. Both devices must use and accept the same protocols in order to understand each other.

Analog vs. Digital

The terms *analog* and *digital* come into play in the transmission of telecommunications signals.

Some telecommunications signals are sent in an analog form, which is represented by a sine wave. The analog signal implies a continuous flow on the line.

A digital signal is discrete, representing the presence or absence of current on the line. Samples of the analog signal can be converted to binary numbers that can then be transmitted in a digital form.

You may hear that a circuit is digital or analog. These terms do not really apply to the circuit, but rather to the ability to transmit the form in which information is sent over the circuit.

Your basic telephone at home has two wires, completing the circuit or loop back to the central office of the local telephone company. The incoming voice and your voice are both carried by this loop. Your voice typically travels over the loop in an analog form. Most telecommunications equipment is designed around this elemental concept.

If you have a PC with a modem, you can connect it to this same telephone line. Since the computer is sending out information in binary form as either a zero or a one, the purpose of the modem is to convert these zeroes and ones into audible tones. These tones travel over the line until they reach the other end where another compatible modem converts them back into ones and zeroes.

We gave some examples of circuits designed to carry digital transmissions earlier in this chapter. These circuits with digital capacity can be used to send voice communications as well as data, but in some cases it is necessary to convert the analog voice signal into a digital form.

This is done with a variety of sampling techniques, but the basic concept is the same. The analog signal (sine wave) is sampled at frequent intervals and given a numeric value based upon the amplitude (height) of the wave at each point. This number is then converted to a binary number. Binary numbers represent all values with only ones and zeros. Thus, the numeric value of each point sampled on the analog signal is converted to ones and zeros which travel across the circuit digitally.

Another issue relating to transmission is whether the circuit is a *2-wire* or a *4-wire* circuit. Most dedicated point-to-point data communications circuits are 4-wire, having a transmit pair and a receive pair. Data can also be transmitted on a 2-wire dial up circuit, usually at a lower speed. On many circuits, the signals travel part of the way on fiber optic cable, but still terminates on copper wire at either end.

The telephone in your home uses a 2-wire circuit.

When ordering services and trying to get two devices to communicate, make sure you clearly state your expectations of how your system will work. Confirm this in writing with the carrier supplying the circuit and with any company supplying hardware.

PART IV

Interactive Systems

CHAPTER 10
Interactive Voice Response (IVR)

What is IVR?

Interactive Voice Response (IVR) refers to the use of a touch-tone telephone to request information from a computer database. The touch-tone signals (sounds) are converted to digital signals understood by the computer. In turn, the digital signals coming back from the computer are converted into a voice that speaks the requested information. The idea is to eliminate the need for a live person to give out information. A well-designed IVR enables you to reach a live person if you need one.

For example, most banks now offer you the capability to dial into a telephone number and enter your account number and a password from your touch-tone telephone. This enables you to obtain information such as your account balance and the checks that have cleared. You may also instruct the computer to transfer funds from one account to another.

Another example of IVR is encountered when you call to request railroad train schedule information. The automated system, using Interactive Voice Response will prompt you with instructions such as the following: "If you're traveling today, press 1; for schedule information tomorrow, press 2; if you're leaving from Grand Central Station, press 1; if you're going to Grand Central Station, press 2; enter the first four letters of the station you're leaving from, using the buttons on your touch-tone telephone; enter the time of day you wish to leave, followed by AM or PM." Often you will be instructed to 'Press 1 if this is correct,' after the system has repeated your selections back to you. The system will then speak to you, providing the information you have requested. IVR is also widely used in Call Centers (See Chapter 5).

Another way to define IVR is by its basic purpose, to use the telephone as an interface with a computer to receive or input data. Although not all IVR applications manipulate data, most receive and disseminate it.

Some IVR systems enable the caller to speak responses to the IVR questions, rather than using touch-tone signals. This is possible due to what is called *speech recognition.* The system might say "press 1 or say 'sales' " (for the sales department).

The term IVR sometimes refers to the application, but also refers to the system itself, which is usually PC- or server-based. The system is also sometimes called a *VRU (Voice Response Unit)*, particularly the part that provides the communications between the touch-tone signals and the computer. It is anticipated that the most popular PC operating system for IVR will be Windows NT due to the fact that it is becoming the corporate platform of choice and gaining the largest market share. Windows NT and UNIX are both used for IVR systems due to their capabilities for multitasking (several things happening simultaneously).

The IVR components reside on circuit boards installed inside a PC, using the PC processing and information storage capabilities. Some IVRs are separately built systems with a proprietary cabinet including circuit boards, processing capability and information storage space.

The trend is toward standardized hardware, giving the IVRs more open architecture and lowering the cost.

The IVR may work in conjunction with a telephone system, perhaps using an Automated Attendant, or callers may dial directly into the IVR on outside telephone lines. It is typically interfaced with a telephone system, enabling the IVR users to escape to a live operator, if necessary, by pressing "0."

IVR and The Client/Server Model

Linking computers together is best defined by the model called *client/server*. The *client* is the screen that you are looking at or the terminal where you are receiving the information (such as your telephone). The *server* is the place from which the information you are viewing or hearing is coming.

The term *host* is sometimes used to refer to a mainframe computer. The context in which we are using it here is to refer to the computer where the information for the IVR application resides. A better term might be *IVR server*.

The expression *PBX-to-host* refers to the capabilities of a PBX to interface with a computer. IVRs are more likely to use a LAN (Local Area Network) and often occupy a dedicated server on the LAN. In both the older mainframe applications and on the LAN, the system works as follows.

How The IVR Provides Information

A telephone call comes through the PBX and reaches an announcement asking the caller to enter or speak his five-digit account code. Or, the call may carry an ANI or Caller ID. The PBX then repeats the information to the IVR and keeps the caller connected to the record located in the computer database. The caller then hears responses relating to his account through the IVR.

If the caller wants to reach a live person, the PBX-to-host capability keeps the computer record attached to the call. The caller and the record arrive at the workstation of the customer service representative simultaneously. The call will be on the telephone and the record will be on the screen of a terminal or PC.

There are two possible locations for data storage in an IVR application. One is on the IVR itself, as a local database. The other is on a separate *host machine*, known as a host database that may be a PC, a mainframe computer (still used by many banks) or, more likely, a server on the LAN. Each of these methods has advantages and disadvantages.

The simplest method of accessing data for an IVR application involves using data that is resident on the IVR machine. The files containing the data are stored on the hard drive of the IVR. Any information that the caller wants to access is already on the hard drive. Information input by the caller is then stored on the hard drive. The information may be imported from another machine or exported to another machine for the purpose of updating the local database, but *all is resident on the IVR hard drive during the session*. The updating may be done via the LAN or using a diskette. It can be done as needed, either real-time (as information is changing), hourly, daily, etc.

There are several ways to make this local database information available to the IVR applications.

— **Local database access**: Some IVR packages have *built-in database functions*. Applications can access ASCII files (called *comma delimited*) using built-in commands. This allows the person setting up the IVR to create a file on virtually any computer system. This may be information such as a customer file, a parts file or an area code table. It is imported into the IVR. The file can then be accessed by the IVR and information can be read out or input to this file. The same ASCII files can be exported at a later date and used for information transfer or the generation of reports. (*ASCII = American Standard Code for Information*

Interchange). Software programmers are writing open IVR systems designed to interface with specific database programs. The use of ASCII for this purpose has diminished as a new interface called *DDE* has been developed. Any database that is DDE compliant can communicate with most IVRs. (DDE stands for Dynamic Data Exchange.)

— **On-Board database**: Some IVRs have the ability to exchange information with other programs that are running on the same server. If this is the case, then the IVR must be running on an operating system capable of running multiple applications simultaneously, such as Windows NT or UNIX. Some IVRs can run a local database manager that allows the users of the IVR access to local relational databases for more complex applications.

— **On-Board spreadsheet**: In the same way that some IVRs can exchange information with other database programs running on the same machine, they may also be able to exchange information with spreadsheet programs on the machine. The spreadsheets allow data access and manipulation.

Advantages of using a local database are the following:

— **Lower Development Cost**: An application using a local database will most likely be less expensive to develop. You will not need to program for file access into a host machine.

— **Less Complexity**: Many of the local database applications are less complex and therefore more easily developed, changed and supported.

— **No Interaction With The Host Computer**: Since there is no direct interface with the mainframe or server, there is no need for complex interaction between the IVR and the host.

— **May Be Faster and More Responsive**: Because no complex interface is required, the local database may be faster and more responsive than a host database system.

Disadvantages of the local database:

— **May Not Have Real Time Information Access**: If the information in a local database is being uploaded or downloaded from a server or mainframe, the information on the IVR may not be completely current. Whether or not this is important depends upon the needs of the application for which you are using the system. In some applications, for security reasons, you may not want the information available real-time.

— **May Be Difficult To Administer**: Complex local databases may be difficult to administer. Applications that require access to a large amount of constantly changing information are not good candidates for local database access.

The more traditional method of data access has been to use data that is resident on a server or mainframe. The files containing the required data are stored on some other computer (server, mainframe PC). Any information that the callers need to access during the IVR session needs to be *read* from the host system. There are several ways to make the connection between the host system and the IVR.

Here are some different types of host connectivity. The type used for a particular application is determined by the hardware type of the host system. For example:

- **IBM 3270 Emulation**: This type of emulation will handle most IBM mainframe host applications.

- **Asynchronous (RS-232) Terminal Emulation**: This type of connectivity enables the IVR to emulate an RS-232 terminal. *Note:* These two represent older technology which is being phased out.
- **SQL Database Interface**: Some IVRs support SQL (Structured Query Language) which is a set of standardized commands for accessing IBM's Database Manager, SQL Server, Oracle and many other databases. Access for this type of interface is usually done via a LAN interface.

Advantages to using a LAN - (or WAN - wide area network) based database for the IVR applications are the following:

— **Real Time Information Access**: Mentioned above as a disadvantage to the local data base. If the success of the application is dependent upon the availability of real time (up to the minute) information, the LAN database will be more likely to provide this.

— **Easier Administration**: Again, as mentioned above, applications requiring access to a large amount of constantly changing information work better if the information remains on the host that is designed to manage those changes.

Disadvantages to the LAN (WAN) database include:

— **More Interaction With The LAN and MIS department**: Since they are interfacing directly with the server or mainframe, there is more need for complex interaction between the IVR and the LAN. This also means more interaction between the IVR vendor and the MIS department. A change in the LAN programs must be coordinated with the IVR application in order for the IVR to continue to work smoothly.

— **May Be Slower and Less Responsive**: Some LAN interfaced systems will be slower and less responsive than a local database system.

IVR Management Information Available

Another consideration in setting up this type of application is how to export the information from the IVR machine into another computer. For example, the IVR is often interfaced with a fax server (see Chapter 11). Information gathered from the application can be stored in various files. It can be formatted into a report or sent to ASCII files. Here are some examples of the types of information that the IVR may export:

- Orders placed (including customer numbers, item numbers, quantities, etc.)
- Fax documents requested (for billing in a Fax-on-Demand application – see chapter on interactive fax response).
- The history of how many callers listened to which type of information.
- Number of times callers used the application.
- Number of times callers used a certain choice on the menu of the application.
- Customized reports – IVR programs are becoming increasingly easier to administer to provide customized reports.

The information may be exported from the IVR in a variety of formats that may include:

- **Into Any DDE Compliant Database**.
- **Printed Reports**: This method enables the application to process the required information right in the IVR, format a report, and send it directly from the IVR to a printer.

- **Faxed Reports**: This method enables the application to process the required information within the IVR, format a report, and send it out to a specific fax machine (see Chapter 11 on Fax Server).
- **E-Mailed Reports.**
- **Audio Reporting Mailbox**: Certain variables may be monitored within an application. This information can be stored in a spoken format in a particular Voice Mailbox where it can be retrieved on a regular basis. For example, you may call into your system to find out how many callers used it on the previous day.

IVR Implementation

The ease of accessing information through an IVR is directly related to the ease of accessing the database from a PC. If one needs to go through several different menus and screens to get to the desired information, the IVR application will probably be set up with the same number of menus or levels. If possible, clean up the ease of access to your databases before implementing the IVR. If callers are going to use the application, it is important to keep it as simple as possible.

What are the reasons for implementing an IVR system? Here are a few from *Call Center Magazine* (212-691-8215).

- **Cost**: IVR is more cost effective than people. If well designed, it does not waste time, does not need time off and will answer calls on weekends and holidays without extra pay.
- **Ending Repetitive No-Brainer Calls:** Requests for bank balances, order status and credit limits are the types of calls that IVR can and should handle. The idea is not to totally

eliminate people, but rather to keep them free to handle the more complicated inquiries.

- **24-Hour Customer Service**: The advantage is that callers can get an immediate answer to their questions, 24 hours a day, 7 days a week. During business hours there is no need to wait on hold to obtain basic information.
- **Proof of Transactions**: IVR provides an audit trail for orders and the distribution of information. As Fax-on-Demand continues to develop, this will become more apparent to the customers of companies using IVR. Bank statements, transaction receipts, order confirmations and invoices will be instantly available.

The trend in the IVR industry is toward a dedicated server on a LAN or WAN, open architecture and server-based platforms. These may be supported by trained non-technical personnel. You buy standard non-proprietary hardware and buy the software separately.

Such systems are becoming more common as standardized hardware becomes the norm. Although Fax-on-Demand (Chapter 11) is being presented separately, the trend is toward the integration of IVR and Fax-on-Demand. This allows more information to be delivered to callers, giving them immediate written confirmation of transactions. These are also being integrated with *screen pops* (see Chapter 12 on Computer Telephony). When a caller exits the interactive system to get to a customer service representative, the representative will have the caller and transaction information on the screen in front of him as the call arrives.

Furthering this trend are *platform based systems* on which Voice Mail, Automated Attendant, Interactive Voice Response and Fax Servers are integrated onto the LAN. This makes them accessible to all LAN users. This enables such things as taking a written fax document and attaching it to a spoken message, and then sending both to the Voice Mailbox of everyone using the LAN. This addresses the trend toward what is called *Unified Messaging.* With Unified Messenging, when you come into your office in the morning, your computer screen shows a list of new Voice Mail, e-mail and fax messages, all in the same place (see Chapter 12).

For more information on how to use IVR see 236 Killer Voice Applications by Edwin Margulies (order from 1-800-LIBRARY or www.telecombooks.com).

CHAPTER 11
Fax Servers

What is a Fax Server?

A *fax server* is a server (PC) on your Local Area Network that sends or receives faxes directly to or from workstations on the Local Area Network. This is usually accomplished by introducing a fax server into an existing LAN. If your organization does not have a LAN, you may still implement most of the capabilities described in this chapter on a separate PC. The trend is for organizations to incorporate fax capability into the Local Area Network to make it more accessible to everyone who needs to use it and to give access to information residing on other servers on the network.

The fax server contains *fax boards*, printed circuit boards that process incoming and outgoing fax transmissions. It also has a hard disk drive for storage and memory. There is also software controlling the fax server applications. This is related, but not the same thing,, as simply having the ability to send and receive a fax using your PC, since it is designed for heavier volume and more complexity. As with other telecommunications systems, sizing depends upon the volume of use anticipated and on leaving room for growth and for adding new capabilities. See Chapter 12 for a discussion of Unified Messaging that provides the capability to receive and respond to faxes, e-mail and voice mail messages all on the screen of your PC at the same time.

Depending upon the nature and faxing habits of your organization, there are several ways to cost justify setting up a fax server. It can clearly cut down on staff time spent feeding printed paper documents through the fax machine and calling back and forth to the recipient to be sure the fax was received. A fax server can also reduce telecommunications expenses, both by enabling the sharing of outside telephone lines and by transmitting at faster speeds, reducing the cost of the telephone call made to send the fax.

Fax servers can integrate with other technologies such as *Interactive Voice Response*, described in Chapter 11. A caller may be listening to spoken information and want a hard copy immediately – for example, instructions for putting together a child's bicycle on Christmas Eve. The caller simply enters his fax number and the information is received by his fax machine within moments.

Fax servers are also used in conjunction with Automatic Call Distribution systems (described in Chapter 5.) Callers queued up waiting for a live representative may be given the option of going to a fax server: "Press 2 to use our Automated Fax Service to have a list of upcoming events faxed to you immediately."

Fax Mail

One of the earliest applications introduced was Fax Mail. This enables people who call into a Voice Mail system to hear voice messages to also retrieve waiting faxes, stored electronically in a digital format. Suppose you are traveling for business. You reach your hotel in the evening and call down to the front desk to get the hotel fax number. Then you call into your Voice Mail, listen to your messages and find out you have three faxes waiting. Using touch-tone, you enter the fax number of the hotel. Within a few minutes your faxes will arrive at the hotel.

Fax mail enables system users to receive fax transmissions directly into mailboxes on the system. Users are notified of fax messages in similar manner to notification of Voice Mail messages.

System users have the option of directing the fax messages to a fax machine at their current location or saving them until they are ready to receive them. Soon some systems will be able to read faxes (and e-mail) to the user over the telephone using *text to speech* technology.

Fax mail software, working in conjunction with a fax board, enables the system to receive the faxes and store them on the hard drive until they are ready to be sent. There are two types of fax mail, *normal* and *annotated*:

- **Normal fax mail** enables outside callers to send a fax transparently. These callers send documents to your published personal fax number and the documents are automatically routed to your mailbox. In this application, each mailbox is given a separate *direct inward dial* telephone number (see Chapter 9 for definition). Some systems can detect whether the call is a voice or fax call, so a separate direct dial telephone number for faxes is not necessary.

  The main advantage of normal fax mail is that it is transparent to the caller. It appears to them that they are sending a fax as usual since it will be no different from sending it to a fax machine.

- **Annotated fax mail** enables system users and callers familiar with the system to send fax messages with voice annotations. A well set-up system will prompt callers throughout the process. This allows the sender to combine a voice message with a fax message or to simply identify the fax with a voice annotation.

The following is a sample of an Annotated Fax call:

✆ Caller is answered by an automated message, calling from a fax-equipped telephone and hears:

"Press 2 to send a message."

✆ Caller presses 2.

"Enter the mailbox number of the person to whom your message should be sent."

✆ Caller enters voice mailbox number of fax recipient.

"This message will be sent to Michelle Barricello. Press to start recording."

✆ Caller presses 2 to start recording.

Beep (Recording voice message).

"Hi, Michelle. Here is the fax of last month's sales figures that we talked about."

✆ Caller presses 2 (Finished recording).

✆ Caller presses 0 for message routing options.

"To append a fax message, press 4."

✆ Caller presses 4.

"Press the start button on your fax machine now."

✆ Caller presses the start button on the fax machine.

Annotated fax mail permits users to add spoken information to the fax being sent. It may clarify the document or make it possible to send an older document with the changes or updates spoken, saving the time needed to generate a new document.

Annotated fax mail also enables the system user to identify a particular fax document from the spoken message, rather than just knowing that there are waiting faxes.

To leave the voice annotation at the same time as the fax is sent, the caller must be calling from a fax-equipped telephone .

An annotated fax mail system may be known as a *Fax Message Center.*

Fax-on-Demand

Another type of fax server application is known as *fax text, fax retrieval* or, more typically, *Fax-on-Demand.*

Fax-on-Demand allows callers 24 hour a day access to a library of documents stored on the fax server. Callers use a menu structure to request the documents they want to have faxed to them.

Any type of pre-generated document (price sheets, brochures, technical information, product instructions, etc.) may be made available to callers with this application. Any caller can access the documents in the library without the need for costly human assistance. Fax servers also support *broadcast fax*, the ability to send a single document or group of documents to a list of separate fax recipients.

Documents can be faxed into the system using the Fax-on-Demand software in conjunction with a fax circuit board. A system administrator numbers documents and creates menus so that the caller can request individual documents. The documents may be stored on the same hard drive as Voice Mail messages or this may be set up as a separate system. The system administrator then maintains the Fax-on-Demand Library.

There are two basic ways for a caller to retrieve documents, *Same Call Fax* and *Call Back Fax.*

With *Same Call Fax*, the caller requesting the documents receives the fax during the same call on which the request was made. For this,, the caller must be calling from a fax machine equipped with a telephone. The following is a scenario for Same Call Fax:

✆ Caller picks up the telephone on the fax machine and dials the access number.

"Good afternoon. Thank you for calling Communications Planning & Services, Inc. If you know the extension of the party you are calling, you may dial it now. For access to our Fax Library, press 7."

✆ Caller presses 7.

"If you are a first time user of our Fax Library, press 1 to have an index of available documents faxed to you. If you know the document number you wish to receive, you may enter it now. For an explanation of the operating procedures of the Fax Library, press 2 now."

✆ Caller presses 3021, the index number of the document they wish to receive.

"Press the start button on your fax machine now."

✆ Caller presses the start button on the fax machine.

The Fax-on-Demand system transmits the requested document.

The advantages of Same Call Fax are the following:

— The caller pays the telephone charges for the fax transmission. Since the caller dialed in from this fax telephone, he pays for the fax transmission as well as the cost to request it. This controls the cost on applications where the Fax-on-Demand owner does not want to incur the charges for the faxing of the document.

— The method is simpler to use than Call Back Fax. It requires only that the caller dial the telephone number of the system, make the selection from the menu and press the start button on his fax machine.

— There is complete control over the faxed document for confidential documents (financial, legal or personal information). Since the caller is right at the receiving fax machine, the document will not be lost or subject to casual exposure.

— Since the document is faxed during the same call as the request, the caller has immediate access to the document.

The disadvantages of Same Call Fax include the following:

— The caller must be calling from a fax-equipped telephone or PC. In addition, callers may not have direct access to the fax machine. It may be in another room or inaccessible (a hotel fax machine, for example).

— Since the call has to be placed from the fax machine that will receive the fax, delivery is limited to the same location that requests the documents. You cannot request a document to be sent to another location (such as to your office if you are out of town) or to another person.

— For the Same Call Fax application to work, a fax port must be available during the call made to request the fax. If no fax ports are available in the system, the system will not hold the caller until a port is free, so the caller must try again later.

With Call Back Fax, the caller requesting the documents receives the fax on a second call, not during the original call. Multiple documents can be requested on the same call. The scenario for Call Back Fax is as follows:

✆ Caller picks up any telephone and dials the access number.

"Good afternoon. Thank you for calling Plantco in Redding, California. Visit our wholesale nursery or call us for landscaping. If you know the extension of the party you are calling, you may dial it now. For access to our Fax Library, press 7."

✆ Caller presses 7.

"If you are a first time user of our Fax Library, press 1 to have an index of available documents faxed to you. If you know the document number you wish to receive, you may enter it now."

✆ Caller presses 3021, the index number of the document he wishes to receive.

"Please enter the telephone number of your fax machine, followed by a pound sign."

✆ Caller enters the telephone number of the fax machine where he wishes to have the fax sent, followed by a pound sign (#).

"The number you entered was ... If this is correct, press 1. Otherwise, press 9."

✆ Caller presses 1.

"To identify your fax, enter your extension or telephone number, followed by a pound sign."

✆ Caller enters their extension number or telephone number, followed by a pound sign (# on the touch-tone dial pad).

"The number you entered was ... If this is correct, press 1. Otherwise, press 9."

✆ Caller presses 1.

"Your fax will be delivered shortly."

The Fax-on-Demand transmits the requested document to the fax machine at the telephone number entered. The document should have a cover sheet that identifies the fax recipient as the person with the telephone or extension number that the caller entered when he requested the fax.

Since the calls delivering the fax documents are made *after* the calls that request the documents, the fax call backs can be queued. This way, the available fax ports in the call back system can be fully used. In some systems, Fax-on-Demand takes incoming calls when it is busy even though all the ports for faxing out might be in use. This possibility for delay is a consideration in building the application. As with each Fax Back application, you must consider how many outside lines are available to handle the incoming and outgoing calls. Otherwise, although ports in the system may be free, if all available outside lines are in use, no calls, incoming or outgoing; will be possible.

When designing Call Back Fax applications, remember that now the owner of the Fax-on-Demand pays for the call, rather than the caller, as is the case with Same Call Fax.

Another consideration that may be a drawback of Call Back Fax is, since the documents may queue before being sent, it is possible that a fax will be sent to an unattended fax machine.

This may not be the best approach for faxing confidential documents or for controlling whether or not the fax has reached the intended recipient.

Methods of Document Selection

There are two basic ways of requesting a document from a Fax-on-Demand application:

- Fax Back from Audio (Spoken) Menus
- Fax Library Using Document Number

AUDIO MENUS

The easiest way to support small Fax Libraries is through the use of an Audio Menu system. This system prompts the caller through a list of choices and allows the caller to select one of the documents based upon an audio description. A typical call scenario would go as follows:

✆ Caller dials the telephone number.

> *"Good afternoon. Thank you for calling Telecom Books. If you know the extension of the party you are calling, you may dial it now. For a catalog of our telecommunications and computer telephony books, press 7."*

✆ Caller presses 7.

> *"For a list of books on telephone systems, press 1. For books on transmission and cabling, press 2. For computer telephony books, press 3."*

✆ Caller presses 1.

> *"For books on small systems, press 1. For books on large*

systems over 200 lines, press 2. All other requests, press 0 now and an operator will be with you shortly."

✆ Caller presses 1.

"Please enter the telephone number of your fax machine, followed by a pound sign."

✆ Caller enters the telephone number of the fax machine where he wishes to have the fax sent.

"To identify your fax, enter your extension number or telephone number, followed by a pound sign."

✆ Caller enters the telephone number of the fax machine where she wishes to have the fax sent, followed by a pound sign.

"The number you entered was... [System repeats the entered telephone number]. If this is correct, press 1. Otherwise, press 9."

✆ Caller presses 1.

"Your fax will be delivered shortly."

Fax-on-Demand transmits the requested fax containing the product information to the fax machine at the entered telephone number. The document will have a cover sheet that identifies the recipient of the fax as the person with the extension number or telephone number that the caller enters when requesting the fax.

The use of Audio Menus has advantages and disadvantages:

Advantage:

- **Simple for first-time callers to use**. Callers have no need to know a document's number. Following the audio menus is straightforward.

Disadvantages:

- **Not suited for a large selection of documents**. With a large number of documents, this method will be complicated for end-users. Fax Libraries that contain more than 10 to 15 documents are likely to be awkward to support when using audio menus. Too many choices in a menu, or menus nested deeper than three levels, may confuse and irritate callers.
- **Heavy Maintenance**. If the Fax Library is changed often, this method requires a large amount of maintenance. New audio greetings and call processor definitions need to be created each time changes are made.

DOCUMENT NUMBERS

The most common way to support a Fax Library is through the use of document numbers. Using this method, the Fax-on-Demand system prompts the caller to enter the index number of the document he wishes to receive. The typical call scenario is as follows:

✆ Caller picks up the telephone and dials the Fax Server access number.

"Thank you for calling DIgby4 Group. If you know the extension of the person you are calling, you may dial it now. For tips on managing your telephone systems and services, access our Fax Library by pressing 7."

✆ Caller presses 7.

"If you are a first-time user of our Fax Library, press 1 to have an index of available material faxed to you. If you know the document number you wish to receive, you may enter it now."

✆ Caller presses 1.

"Please make your selection from the following documents:
For Dollar Saving Tips, select 11
For Tips on Purchasing New Telephone System, select 12
For Tips on Training Your Switchboard Operators, select 14
For Tips on Purchasing Local and Long Distance Calling, select 15.
For Tips on Maintenance Agreement Negotiation, select 16.
For Telephone Bill Management Tips, select 17."

✆ Caller presses 16, the index number of the document requested.

"Please enter the telephone number of your fax machine, followed by a pound sign."

✆ Caller enters the telephone number of the fax machine where they wish to have the fax sent, followed by a pound sign.

"The number you entered was... [System repeats the entered telephone number]. If this is correct, press 1. Otherwise, press 9."

✆ Caller presses 1.

"To identify your fax, enter your telephone number, followed by a pound sign."

✆ Caller enters their extension number or telephone number, followed by a pound sign.

"The number you entered was ... [System repeats the extension number]. If this is correct, press 1. Otherwise, press 9."

✆ Caller presses 1.

"Your tips will be faxed to you shortly. Thank you."

Fax-on-Demand transmits the requested document to the fax machine at the entered telephone number. The document will have a cover sheet that identifies the recipient of the fax as the person with the telephone number that the caller enters when requesting the fax.

If the caller did not know the number of the document he wanted, he would press 1 after the first set of prompts. The system would ask for a fax telephone number and then fax him an index of available documents.

ADVANTAGES & DISADVANTAGES – USING DOCUMENT NUMBERS

Advantages:

- **Easy to use**. With a large number of documents, this method is the easiest for end users to use and understand. Even with as many as several hundred documents, the menus remain brief and simple.
- **Light Maintenance**. Even if the Fax Library is changed often, this method requires only a minimal amount of maintenance. No new mailbox greetings or Call Processor changes need to be done in order to add new documents.

Disadvantages:

- **May require first-time callers to call twice**. First-time callers may need to place two calls to use the system; a first call to request the document index and a second call to request the actual document.

Broadcast Fax

The Fax-on-Demand software package may also support a feature known as *Broadcast Fax*. This allows a single document or group of documents to be broadcast to a list of people at different fax machine locations. The system administrator can create a list of fax numbers (including a name and title for the receiving party). He can then send a document or combination of documents to all parties on the list. The fax transmission time and date can be specified, allowing for the scheduling of fax delivery at off-peak times.

CREATING THE FAX LIST

Creating the list of fax machine telephone numbers and recipient names is a very simple process. Using any word processing system, the system administrator creates the list in the following format:

1-XXX-XXX-XXXX, Name

1-XXX-XXX-XXXX, Name

1-XXX-XXX-XXXX is the telephone number of the fax machine, and Name is the recipient's name and/or title. The system administrator then uses a word processing system to export the file as an ASCII text file. This file is then placed into the floppy disk drive of the Fax-on-Demand system. Some cautions:

- The local word processor must be able to exportASCII text files (almost every commercial package has this ability)
- The computer that runs the local word processor must have a 3 1/2" floppy disk drive.
- Only certain Macintosh computers can export files in the correct format. Find out if you can do this if you are a MAC user.

- The system administrator must have physical access to the Fax-on-Demand system in order to load the disk into the floppy drive.

GETTING THE DOCUMENTS INTO THE FAX SERVER

Loading the documents you wish to send with the Broadcast Fax feature into the fax server is done in the same manner as loading the documents for the Call Back Fax application. These documents can be faxed in from a fax machine or created with word processing or desktop publishing software and faxed in from a computer with a fax board.

SCHEDULING THE TRANSMISSION

To schedule the transmission of the Broadcast Fax, the system administrator typically performs the following functions:

a. Loads the disk with the fax list into the fax server floppy drive or retrieves from a hard drive.
b. Goes to the Fax Send Screen on the fax server.
c. Enters the name of the fax list file in the drive.
d. Enters the date and time for the transmission to begin.

Fax Library Management

LIBRARY MAINTENANCE

The maintenance of the Fax Library is straightforward. It is done from the Fax Administration screen of the Fax-on-Demand system. The following functions are supported:

- Receive a document
- Send a document

- Copy a document
- View a document
- Fax Reports

RECEIVE A DOCUMENT

Adding a new document to the library is a simple function. The system administrator enters the Fax Maintenance screen and selects Send or Receive a Document. After specifying which fax port will receive the document and the name of the document, the administrator goes to a fax machine and calls the Fax-on-Demand fax port. The fax port answers and receives the document. The document is now available for use by the Fax-on-Demand system. If the fax server is on a LAN, the documents can be updated from another computer on the LAN.

SEND A DOCUMENT

Any document in the system can be manually sent from the Send or Receive a Document screen. The telephone number of the destination fax machine, the recipient's name, the name of the document, and the time and date you wish the document sent are entered from the *Send a Document* screen.

COPY A DOCUMENT

Sometimes a document is received under a name that simplifies the administration of the Fax-on-Demand application. A price list might be received as prclist.fax. In order to use this document in an application that uses document index numbers, it is necessary to rename it with a name containing only numbers. The copy feature is also used to trim off the fax header for documents faxed into the library. This is easily done from the Fax Document Management screen.

VIEW A DOCUMENT

Documents stored in the Fax Library may be viewed on the system administrator's screen. The resolution may not be such that all documents can be easily read, but the documents can at least be identified in this manner.

FAX REPORTS

The Fax-on-Demand software provides management information such as:

- *Fax Call Log* – A log of all incoming and outgoing calls made on the fax ports.
- *Fax Error Log* – A log of all errors and system start-ups involving the fax server.

DOCUMENT STORAGE

Fax documents are stored as digital files on the hard drive of the fax server.

Fax documents can vary in file size from 50k bytes to 110k bytes per page. As a rule of thumb, the average fax document is 75k bytes. Ten megabytes of drive storage holds roughly one hour of voice storage. Ten megabytes of drive space holds roughly 120 pages of fax storage. A good rule of thumb for calculating storage requirements is: *Each hour of storage holds 120 pages of fax.* Storage methods are continually being improved upon to be more efficient.

Methods of Creating and Importing Documents

FAXING DOCUMENTS

The simplest way to send documents into the Fax Library is to fax them in from a fax machine. The disadvantage of this method is a reduction in image quality. Each time a document is faxed, the image quality deteriorates. Horizontal and vertical lines start to look rough (called *jagging*) and fine detail starts to be lost. Most documents can be faxed into Fax-on-Demand for the initial setup without a serious amount of image deterioration. It is important to start that process with a clean, sharp original.

WORD PROCESSING

Fax-on-Demand can also import documents from most word processors. Any word processor that can export a document in the ASCII text mode (Word, WordPerfect, etc.) can be used to generate documents for a Fax-on-Demand application. These documents are imported into the fax library by transferring them on a floppy disk or over the LAN to the fax server.

PC FAX BOARD

The most efficient, professional way to create a Fax Library is to use a PC fax board. These are commercially available boards that fit into a slot inside a personal computer and allow that computer to emulate certain fax functions.

Most PC fax cards can:

- Receive a fax.
- Send a fax.

- Convert received faxes into graphic files (which can be edited and cleaned).
- Convert special files (Postscript, etc.) into faxes.

These boards can be used in a PC running a desktop publishing system to create and load professional quality documents into a Fax-on-Demand application.

Fax-on-Demand and Fax Broadcast Applications

Certain departments and functions that are found in almost every business can benefit from the Fax-on-Demand applications including:

Human Resources or Personnel:

- Job postings
- Payroll information
- Holiday schedules
- Benefits explanations
- Job descriptions
- Hiring application forms
- Company procedures

Accounting, Payroll and Finance:

- Expense forms
- Payroll forms
- Capital investment or asset acquisition forms

Outside Sales:

- Product brochures
- Information sheets
- Order forms
- Price sheets
- Contact documents
- Configuration guides

Inside Sales or Order Desk:
- Product brochures
- Specification sheets
- Price sheets
- Order forms

Telemarketing:
- Product brochures
- Specification sheets
- Company information sheets

Technical Support:
- Technical tips
- Product information
- Programming information or forms
- Wiring diagrams

Warehouse or Logistics:
- Product sheets
- Order forms

Marketing or Support:
- Product brochures
- Program information
- Price sheets

Training:
- Class schedules
- Class descriptions
- Class prerequisites
- Class locations

Field Technical Staff:
- Technical tips
- Programming forms
- Service forms
- Training documents

Fax-on-Demand Cost Justifications

There are as many justifications for the Fax-on-Demand as there are applications. Each user will place a slightly different value on it depending on the specific business and applications. Below are a few of the common justifications for Fax-on-Demand.

Increased Service and Value to the Company Using Fax-on-Demand:

- 24 hour access to information
- Ease of access to information
- Remote access to information

Increased Efficiency of Employees:

- 24 hour access to information
- Ease of access to information
- Remote access to information

Administration Time Saved:

- Employee time on the phone
- Employee time at the fax machine
- Waiting for an available fax machine
- Searching for printed documents
- Time spent mailing out documents and bulletins
- Time spent explaining technical documents over the telephone

Reduced Mailing or Express Service Charges:

- Money spent on sending out price sheets, etc.
- Money spent on sending out brochures or fact sheets
- Money spent on sending out sales promotional literature
- Money spent on sending out technical documents

Simplified Document Administration:

- Price lists
- Technical information
- Business forms

Configuring Fax-on-Demand

The configuration of Fax-on-Demand can be broken into three areas:

- Hard Drive Requirements
- Fax-on-Demand Hardware Requirements
- Telephone System Requirements

Each of these areas has its own set of configuration requirements.

HARD DRIVE REQUIREMENTS

Hard drive space is needed to hold the documents of the Fax-on-Demand Library. This additional space must be taken into account when the basic system is sized. By following the basic formula of 1 Hour = 120 pages of Fax-on-Demand (an industry average) you can determine the additional hard drive capacity required to support the application. These applications always have a tendency to grow as new uses for the module are uncovered. Be sure to allow some extra hard drive capacity for growth.

HARDWARE REQUIREMENTS

The Fax-on-Demand application uses fax boards for the fax server. Each fax card requires a slot.

TELEPHONE SYSTEM REQUIREMENTS

Each fax port on the fax server system needs either a separate central office line or a separate analog port from the telephone system. The choice between outside line and analog station port is controlled by the application:

— Applications using the Same Call Fax method must have dedicated analog telephone ports for each fax port. Many Fax-on-Demand systems now use a combination of DID (direct inward dial) for incoming and combination trunks for outgoing calls.

— Applications using Call Back Fax may use either a dedicated analog telephone port or a dedicated outside line for every fax port. The outside lines can come directly into the fax server from the telephone company or a long distance company or can be connected through the PBX on-site.

On different telephone systems, different hardware is required to create analog telephone set ports. Considering application and telephone system type, it may be more economical to use separate outside lines. See Chapter 9 for more information on types of outside lines.

When purchasing a Fax-on-Demand system, here are some of the questions a knowledgeable vendor will ask you.

Questions about your business:

- What is your business and what types of activities do you conduct?
- How many departments are there and how are faxes sent and received in each?
- In what instances do customers call asking the same questions repeatedly?
- Do you have a separate customer service department?
- What type of telephone system do you have?

Fax-on-Demand Specific Questions:

- How many documents are there to be retrieved?
- How frequently will they be retrieved?
- How time sensitive are they and how frequently do they need to be updated?
- Will your application be serving different time zones?
- What overall volume are you anticipating? Including:
  - Number of pages stored
  - Number of callers
  - Number of simultaneous callers
- What capability for growth is anticipated?

When you buy a Fax-on-Demand system, find out what the upgrade costs will be and how many fax ports can be added.

It is also a good idea to check the quality of the documents generated by the system you are considering.

Other capabilities of fax servers to look for when you are shopping for one include the following:

— Least cost routing (sending the call over the lowest cost outside line).

— Client/server architecture, providing all processing at the server level, freeing individual workstations to work in other applications.

— Ability to interface with other applications you may be using.

— Flexible licensing arrangements as the number of users increases.

📖 For more information on Interactive Fax Response, see Computer-Based Fax Processing by Maury Kauffman (order from 1-800-LIBRARY or www.telecombooks.com).

PART V

Computer Telephony

CHAPTER 12
Computer Telephony

Computer Telephony is a term coined by Harry Newton of Newton's Telecom Dictionary fame. The concept of Computer Telephony, or *CT* as it is called, covers a developing array of hardware, software and applications that add computer intelligence to the making and receiving of telephone calls.

The way in which CT continues to develop will be shaped by the technology, the tenacity of the companies involved and the demands of the marketplace. The types of systems and applications described here are manifestations of the current state of Computer Telephony, sure to change.

You may also hear the expression *CTI* or *Computer Telephony Integration.* CTI refers to the integration or merging of different systems (hardware and software) into a Computer Telephony application designed for a specific purpose. There is a trend toward building systems incorporating many capabilities (such as PBX, Voice Mail, Automated Attendant, IVR and Fax-on-Demand) into one system and all on one software platform. This is beginning to diminish the amount of integration activity required in setting up a system, although integrating with a computer database can still be time-consuming and complex. In some cases, if an organization has a large investment in a system, such as an IVR for example, having spent many hours and dollars setting it up, they may opt for integration rather than starting over with a new IVR, even though it may be a built-in capability of the system they purchase.

As with all technology purchases, organizations wanting to roll out Computer Telephony applications must consider the systems and procedures in place and the costs and trade offs of different options.

Computer Telephony has manifested itself in the following types of systems and capabilities:

IVR

We described IVR (Interactive Voice Response) in Chapter 10. IVR is an example of Computer Telephony, enabling telephone callers, using touch-tone signals or speech, to access computer-based information over telephone lines.

Fax Server

Fax servers supporting applications such as Fax-on-Demand also represent a CT type of system. Documents stored in a computer can be faxed to callers who dial in and request them, using a touch-tone telephone or speech. This is covered in Chapter 11.

Screen Pop

A Screen Pop refers to the delivery of a computer screen of information about a caller that arrives at the desktop at the same time as the telephone call. Callers may be recognized either by using touch-tones or speech to identify themselves to an IVR system answering the call. "Thank you for calling Comware Systems, Inc. For immediate hotline support, please enter your 5 digit customer identification number." Callers may also be identified by the number they're calling from (Caller ID or ANI) or by the telephone number they dial into (DID or DNIS). Once the caller is identified, the telephone system sends a message to the Local Area Network server identifying the caller and instructing it as to which desktop the screen of information is to be delivered.

Unified Messaging

Unified Messaging is an application that attempts to streamline the work of receiving and responding to multiple messages from voice mail, e-mail and fax. It has been developed primarily by Voice Mail companies as an expansion of their system's capabilities. A user of unified messaging receives e-mail, voice mail and fax messages all on the same software platform. And all appear on the computer screen at the same time and can be scrolled through, responded to or deleted. The Voice Mail is heard through the computer. With Unified Messaging, laptop users can be listening to and responding to Voice Mail messages from anywhere, without being connected to the Voice Mail system. When they log onto the system with their laptop, the Voice Mail responses they've left will be sent to the recipients.

Placing Calls From The Computer Database

Many individuals in the workplace have a software database or *PIM* (*personal information manager*) of names, addresses, telephone numbers and other contact information on the PC. These programs enable you to dial out by using the computer mouse to point and click on the telephone number on the screen, rather than having to dial the telephone. This is one of the less complex CT applications and may be viewed as computer based *speed dialing.*

IP Telephony

IP Telephony is a developing subset of Computer Telephony and is often presented as a separate discipline altogether.

IP stands for *Internet Protocol.* This is a standard set of instructions for communicating among computers that was agreed upon by a group of universities in the 1960's. Supported by the

U.S. Military for research and development, the original network was called *ARPANET*. They set up a network among themselves to exchange information over telecommunications circuits. This was the was the foundation of today's Internet supporting the *World Wide Web*.

The term *VoIP* refers to voice communications using Internet Protocol. Most telephone calls are currently made over what is sometimes called the *PSTN* or *Public Switched Telephone Network*, the network of circuits connected by the switching systems of the many local and long distance telephone companies worldwide. A voice conversation on the PSTN requires an end to end connection which stays in place and uses up bandwidth on the circuit for the duration of the telephone call. During the call, the voice may be converted from analog to digital form and back to analog at the distant end.

It is also possible to transmit and receive a voice conversation by breaking the voice transmission up into digital *packets* (groups of zeroes and ones) and sending it over a circuit in that form. In this scenario, the end to end connection used exclusively for the voice conversation on the PSTN is not required. More efficient use of circuit bandwidth is possible by sending packets of other types of transmissions such as computer data, interleaved with the packets of voice. IP is the protocol used to enable this.

Some large corporations who have their own wide area networks in place connecting multiple locations are using VoIP to send voice communications over the same network that the data communications travels over.

While, technically speaking, VoIP can also be used for voice conversations over the Internet, this is not practical yet since the volume of communications on the Internet is uncontrolled by any one entity and voice calls can be delayed to a point where a conversation would take too long or not be clear. Technology is rapidly being developed to address this problem which will cut down on the effect of these delays.

Communications Servers

There is a developing group of telephone systems sometimes called *communications servers*, implying that voice communications is becoming simply another function of the computer network. Other terms used to describe these systems are *UN-PBX* and *communications appliance.*

As with any developing technology, different systems are being put together in different ways. Some of these have separate telephones while others present the telephone capabilities (transfer, conference, etc.) on the computer screen, using the computer mouse to control telephone functions.

Some communications servers send the voice over the same pair of wires as the data communications and may use IP or some other communications protocol to accomplish this.

Some are based upon Windows NT which simplifies the task of administering the telephone system since it can be done by someone who also administers other NT systems on the computer network.

History of Computers and Telephones Merging

Here are some of the former concepts of "Computers and Telephones coming together" which are *not* what is meant by Computer Telephony:

1. Digital PBXs are specialized types of computers designed to switch telephone calls. They are made up of printed

circuit boards and have programmable software and memory; therefore, they really are special purpose computers.

2. Digital PBXs can now serve as the vehicles for switching data communications as well as voice communications. In the early 1980's, the PBX manufacturers were offering their products as vehicles for connecting computers in a local area network. Separate circuit boards for *data ports* were sold. Some manufacturers added data ports on the same circuit board as the voice ports. Telephone instruments were made with *RS-232* openings in the back for plugging in the computer, or the telephones sat on top of separate data devices into which the computer would be connected. This whole concept never took hold for a variety of reasons. The three main ones were probably (a) It added significantly to the cost of the PBX and therefore to the cost of installing the local area network; (b) There was resistance to the idea of "putting all your eggs in one basket" by switching everything through the PBX; (c) The MIS managers did not talk to the telecommunications managers and certainly did not want their data switched by the PBX in the telecommunications department's domain.

3. Voice and data communications frequently travel over the same cable (different pairs of wires) or over the same high-capacity circuit (different channels).

One significant difference in today's Computer Telephony is that this time it is driven by the computer industry rather than the telephone industry.

Erector Set Computer Telephony

Computer Telephony Magazine coined the term Erector Set Computer Telephony in 1994. The Erector Set metaphor is still appropriate to describe much of Computer Telephony today. An Erector Set has many pieces, large and small, which can be put together in a variety of different ways limited only by the imagination of the builder. Another aspect of an Erector Set is that sometimes the end result does not look the same as you thought it would when you began to put it together.

There are many components, both hardware and software, being made by a variety of companies which can be put together in different ways for Computer Telephony applications.

Some forces at work driving Computer Telephony include:

1. Many large companies in the computer industries, including Microsoft, have entered the business. Windows Telephony enables you to use your PC as a telephone.
2. The telecommunications industry has shifted. Telephone system manufacturers were once preoccupied with keeping their products proprietary with a "closed" architecture for both hardware and software. They are now touting the "openness" of their systems and developing them to be even more open. They are also forming alliances with other companies in the Computer Telephony field to enhance the capabilities of the telephone systems. We may see the traditional large PBX vendors buying up the companies making the new communications servers or coming up with their own version of this.
3. Technology has progressed to a point where Computer Telephony applications can be made to operate smoothly. The new powerful *digital signal processors* (DSPs) are a driving force.
4. There are many new Computer Telephony standards making it easier for companies to purchase different components which work together.

5. Business people are quickly becoming aware that they can create customized telecommunications tools just as they have done with customized computing tools. They no longer need to be restricted by the rules set forth by a telephone system manufacturer.
6. The *Internet* has some capability for basic telephony.

The idea behind Erector Set Computer Telephony is that the PC, either as a standalone or as part of a local area network, adds intelligence to the telephone and the telephone network. The telephone industry is substantially larger in size than the computer industry, so the opportunities to change existing telephone networks are significant.

These opportunities include:

1. **The Open Desktop** – The idea behind this is to redefine the telephone that now sits on top of your desk next to your PC. The telephone functions can now be controlled by your PC in ways that are easier to use and that integrate the telephone calls with information in the PC. The PC can actually become the telephone, with all of the telephone functions on the screen and the handset hanging on the side of the display monitor. The addition of a circuit board can give existing PC's the capability to act as a telephone. Most of these applications use Microsoft's *Windows Telephony.*

   Some telephone system manufacturers, in an effort to remain competitive and preserve the viability of their proprietary telephones, have developed their own software to run on a PC in conjunction with the telephone. This software provides functions such as the maintenance of a telephone directory and automatic dialing. The approach of these manufacturers is to focus on the processing of the telephone call: answering, putting on hold, transferring, conferencing.

The Microsoft approach includes call processing, but expands to incorporate voice processing. It refers to such capabilities as reviewing Voice Mail messages and appending Voice Mail messages to documents that may then be forwarded to someone else.

There are enormous personal productivity benefits to be realized by running your office telephone from your PC. Your telephone rings and a screen pops up with your notes from your last conversation with the person who is calling.

You pull down the names of people whom you want to conference and then automatically fax them an agenda and arrange for the conference call to be set up at a certain time.

Once you are able to do things like this, there will be no turning back to the old time consuming methods of working.

2. **Open Local Area Network (LAN)** – This can be thought of as an extension of the open desktop, except that the telephony functions need only be put into the LAN server rather than into every PC on the network. The circuit boards in the LAN server talk to your office telephone system, your Voice Mail system, your electronic mail system, etc.
3. **Open PC Voice Processor** – Voice processing includes Automated Attendant, Voice Mail and interactive voice and fax response. All of these functions can be provided by a PC which can be connected to other systems with which information is exchanged and shared, including the telephone system and the local area network.

4. **Open Toolkits** – These are software kits enabling those with some programming ability to write applications for interfacing and integrating the hardware and software components of Computer Telephony. Many of the toolkits are provided at no charge by the manufacturers in the hope that those developing applications to work with their existing products will increase the demand for those products.

5. **The Dumb Switch** – There are do-it-yourself switches (PBXs) that come without any predesigned software and need to be programmed from the ground up. These differ from most PBXs currently being used by businesses, which come with proprietary software enabling you to have a minimal amount of customization. They are sometimes part of the group of systems called the UN-PBXs.

   Some dumb PBXs are designed for processing telephone calls only. They can be used in a variety of specialized applications, including switching of cellular telephone calls. Most PBXs from the major manufacturers have not taken advantage of the opportunity to customize the systems to fit particular types of businesses, other than perhaps the hotel/motel industry.

   Customization to complement your business is what Computer Telephony is all about.

   Some of these dumb switches are designed to conform to current voice processing standards. They come equipped with interfaces to other voice processing circuit boards for Voice Mail, fax, speech recognition, voice recognition, speech-to-text, etc.

6. **Open Long Distance Networks** – The concept is to view the network not as a network but as a platform for developers. With that assumption, the primary tool for

developers is Signaling System 7, now ubiquitous in all long distance telephone companies and increasingly in local telephone companies.

All of the major long distance companies now view their network as a platform for development of specialized applications and are evaluating the best ways in which to develop and deliver services. These carriers are also planning to roll out VoIP (voice communications using Internet Protocol) to make more efficient use of their networks in place.

Some carriers have released specifications that will enable customers to interface with the network to control their 800 calls in a variety of ways and to obtain information on these calls. For example, the customer can have a database which blocks calls from any customers who have bad credit. Why pay for a call from someone with whom you do not wish to do business?

7. **The Computer Equipped For Telephony Capability** – Computer manufacturers are now equipping their computers with telephony features to enable them to interact with telephone systems, Signaling System 7 on networks and other telecommunications applications.

Windows Telephony

In May of 1993, Microsoft and Intel announced *Windows Telephony.* This included a telephony interface standard for application developers, hardware manufacturers and service providers.

The specifications of this interface are available, as is a Software Developer's Kit. This will give software developers the tools needed to create the off-the-shelf software, and the hardware developers the tools to create the necessary capabilities for implementation.

Windows Telephony is basically a set of functions within Windows that enable a call to be placed over the telecommunications network. It allows any application written to Windows to tap the resources of switched telephone networks. The goal is to get rid of the bottleneck in bringing the power of the PC to telephony. There are several factors currently contributing to this bottleneck:

1. It is difficult to interface the variety of telecommunications switches that are in use both in the central offices, long distance carrier sites and at business locations. For example, no manufacturer's PBX will work with another manufacturer's telephone.
2. It is difficult for any software to communicate with the various telephone switches.
3. Most business telephone systems do not bring the local telephone company dial tone or its equivalent (an analog telephone extension) to the desktop. Computer programs currently in use for functions such as rolodex and automatic dialing need this dial tone to operate. They enable the caller to pick up the telephone handset and talk once the number has been dialed. To get around this, some business people have a separate telephone and telephone line at the desk for using computer applications. This is typically not connected to the PBX system, for which there is a second telephone!

The intent of the Windows interface is to eliminate these problems.

There are three elements to Windows Telephony:

1. **The API (Applications Programming Interface)**. This is a standard specification to which a telephony applications developer must adhere.

2. **The Windows DLL**. This is software code that allows Windows to manage the applications in conjunction with the hardware and network services.
3. **The SPI (Service Provider Interface)**. This is the interface between Windows and the telephony hardware. It may either be a proprietary telephone on a PC circuit board or a box interfacing to cellular, ISDN or POTS (plain old telephone service).

Another promising application is screen-based telephony, which makes the telephone system functions easier to use through Windows. For example, the average business person does not know how to use the telephone to set up a simple three person conference call. Screen based telephony enables you to "drag and drop" three different names into the conference. The PC will look up the telephone numbers, dial them and establish the conference for you.

Additional up to date information on the Windows Telephony interfaces can be found on the internet at *www.microsoft.com.*

Telephone Call Control

Some of this information is from a booklet on Computer Telephony published by Mitel (800-267-6244). Thanks to Lou Kratzer and Bill Cibbarelli of Mitel for providing it.

The starting point for Computer Telephony development is a set of basic services including Call Control, Call Monitoring and System Feature Activation. The technology enabling Computer Telephony is software that communicates with the telephone system, telling it what to do. This application can access a set of commands such as Make Call, Answer Call and Transfer Call. When a command is issued by the Computer Telephony software, the telephone system attempts to complete the assigned task and reports back to the application with the result. The result might be complete success (the call went though), progress has been made (the other end is now ringing) or failure (the dialed number is busy or was not answered.)

This information must be provided on a real time basis, as the events occur. The application design has to allow for real life situations – for example, peak times when all the outside lines are busy, or "power users" who switch back and forth between several calls on hold. Users have come to expect almost instantaneous response from their telephone systems. The Computer Telephony application designer must deliver based on that expectation.

Call control expects the application to act as if it were a telephone set. Anything that the telephone can do, the application can now do, too. There are two approaches to Call Control:

First Party Call Control

The basic premise of First Party Call Control is that the Computer Telephony application is acting on behalf of one user. The application runs on the user's desktop PC. There is a physical connection between the application, the user's PC and the user's telephone line.

The PC may connect in front of the telephone, behind the telephone or may actually replace the telephone (see Figure 12.1). The point is the application is acting on behalf of *this one user only*. Through the application, the user controls the telephone call. Examples of this include Personal Directories, Personal Answering Machine and Personal Call Accounting (tracking calls made and received).

Figure 12.1

First Party Call Control

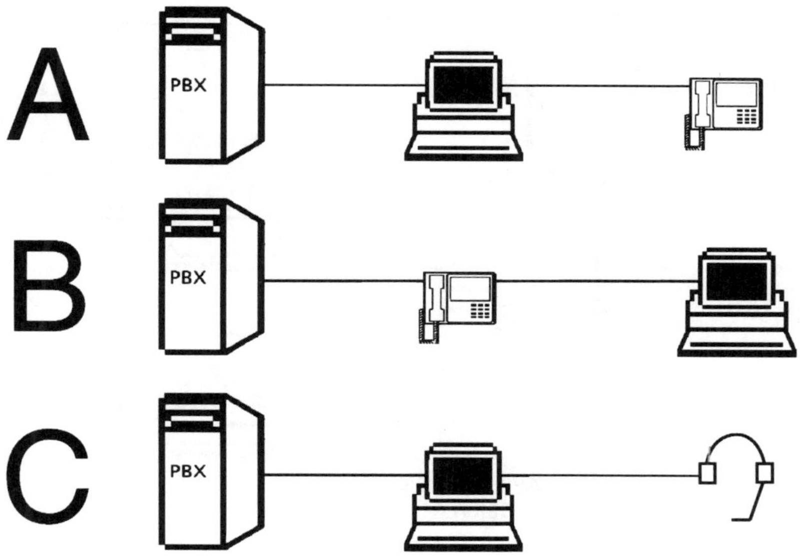

Compliments of Mitel

The accepted standard for first party call control is called *TAPI* (*Telephony Application Programmer's Interface*) and is the basis for Windows Telephony.

Third Party Call Control

The premise of Third Party Call Control is that the Computer Telephony application acts on behalf on any of the PCs (clients as part of the client/server model) that are part of a workgroup using a Local Area Network (see Figure 12.2). The application runs in a shared server. There is no direct physical connection between the user's PC and the user's telephone line. Instead, there is a logical connection – the PC application talks to the server, which in turn

Figure 12.2

Third Party Call Control

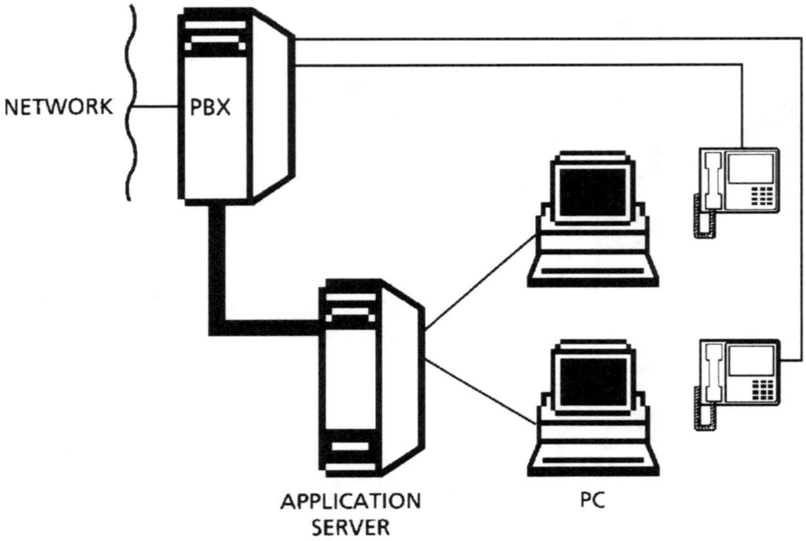

Compliments of Mitel

controls the PBX. The server is acting on behalf of the user. "Make a call for extension 113" or "Answer the call for extension 205" are examples of Call Control commands available. The shared server can offer both personal and workgroup services such as personal and workgroup level directories, organizers, etc.

The server provides a coordination point for all calls being handled in the workgroup. This makes a much more powerful level of call control possible than the First Party Call Control. The central server-based application can handle the distribution of all calls to the members of the workgroup, including activities like Call Screening or Backup Call Answering.

The standard for Third Party Call Control is *TSAPI* (*Telephony Services Application Programmer's Interface*), which was developed by Novell and AT&T.

Call Monitoring

Both types of Call Control expect the application to act like a telephone. There are other capabilities inherent in the PC, such as Call Monitoring. The application can set a Call Monitor in the PBX to collect information on almost any activity. For example, by setting a monitor on a single user's telephone set, the application can watch every button pushed, every digit dialed and the picking up and replacing of the handset.

By monitoring any trunk (outside line), the application can see each incoming call, collect information coming in such as the calling number (*ANI*) or the number dialed (*DNIS*), watch to where the call was directed and know when and where it was answered.

By selectively monitoring telephones, groups of telephones or trunks, the application can get as detailed a picture of the PBX activities as required to make the application work. This is especially valuable in generating management reporting and performance measuring statistics.

Feature Activation

Most PBXs provide over 200 features to improve call handling. The majority of users never use more than four of them! The use of PC-based applications that can be set to the user's preferences unlocks the capabilities built into the telephone system. These applications allow simple, screen-based control of features that are otherwise cumbersome to access.

Using a mouse to point and click on your computer screen, you can activate and deactivate features in the telephone system. For example, a Personal Organizer application could set up *call forwarding* for a user who is away from the office and turn off that call forwarding when the user returns.

In another example, a Computer Telephony application could modify Call Screening by a secretary on behalf of a workgroup. The screening would be turned off at the end of the day and calls would be redirected to an answering service.

Two Approaches to Computer Telephony Integration

Switch Links

This category of early Computer Telephony implementations was accompanied by many technical names for the link or interface between the PBX and the computer. These includes OAI (Open Application Interface), Call Path, Passageway, and Meridien Link, to name a few. What these all had in common was an architecture rooted in the "mainframe computer to mainframe PBX era." These were complex implementations made largely through alliances between big PBX manufacturers and big computer manufacturers. The solution was an enterprise (company-wide) solution.

Communications Server

The newer approach uses a *communications server* which is installed as another node on the Local Area Network. This server is equipped with the hardware and software elements necessary to deliver Computer Telephony solutions to that workgroup. The telecom server connects directly to the public network to handle all calls coming into the group and connects directly to the desktop client to deliver those calls. This approach allows all the priority customer contacts to bypass the enterprise PBX in the basement. As a result, the enterprise PBX no longer needs to be upgraded. The server has a few simple connections back to the

PBX to allow internal calls between the workgroup and the rest of the organization. A variation of this is for the telephony server to connect to the PBX rather than directly to the public network only for the purpose of accessing outside lines.

Communications servers start out as basic computers with new circuit boards added for the Computer Telephony applications.

There are four categories of these circuit boards:

- **Digital trunk board** – Provides connection to advanced network services such as ISDN primary rate interface.
- **Analog trunk board** – Connects internal voice calls to the company PBX if the telephony server connects directly to outside lines.
- **Digital Line Board** – This passes the voice or video connections to the desktop.
- **Special Purpose Resource Boards** – For specific technologies such as video compression or speech recognition.

Note: The traditional PC is not designed to handle the large bandwidth required to transport real-time voice and video information (e.g. from the digital trunk card across to the digital line card or to a voice processing card). To handle this, a secondary telecom bus (a circuit board) is added to the server architecture. There are two competing standards for this telecom bus within the industry – *SCSA* from Dialogic and *MVIP* from Natural Microsystems. Boards are available for each standard from numerous vendors.

Unified Messaging

The idea behind *Unified Messaging* is the consolidation and streamlining of the variety of methods through which we now receive communications. First you check your Voice Mail, then you check e-mail. You receive faxes throughout the day and a stack of paper mail appears on your desk every morning.

Integrated messaging brings all of this together onto your PC screen. You can browse through your Voice Mail messages, see who they are from, listen to them, store them or discard them. You can do the same thing with your faxes, electronic mail and paper mail (scanned into the computer network) and actually see the documents displayed before you. You can act on each, respond, store or discard or add your own notes or voice message and send the whole thing to someone else for action.

Unified messaging provides you with the opportunity to sort out the many pieces of information and communications reaching your desk every day and TO assign priorities to them.

Unified messaging can also help you to respond to incoming calls. It lets you know who is calling and gives you the option of answering the call, answering the call after a slight delay where the caller will hear a delay message ("This is Michael Berezein; I will be with you in just a moment."), or rerouting the call to Voice Mail or to someone else in your office.

Unified messaging has been described as "your very own electronic personal secretary" in the form of a desktop Graphical User Interface (GUI). It accesses a communications server on a Local Area Network (and the data therein). It controls your telephone, lays out all of your voice/fax/electronic/image mail on your PC screen. It integrates everything with other programs spreadsheets, word processors, etc. most likely in a Windows environment. The ingredients: a Local Area Network, a voice processing server that includes various communications-medium resource hardware (voice circuit board, fax circuit board, etc.) that hooks into electronic mail and closely integrates with the PBX and Windows- based client software running at the desktops.

The key to this type of integrated multi-media messaging is that it enables people to decide in real-time how to handle calls and do it politely. It is as if your own very efficient personal secretary were handling it. It is intelligent disposition of information.

The intelligence results because the message handling choices are easily accessible that you can make decisions on the run. Your choices are visually arrayed so that you may absorb them quickly and click on the right choice.

Using the telephone alone, without the screen, you must listen to the voice reading all your options (Press 1 to route the caller to Voice Mail, press 2 to put the caller on hold, press 3 to send the caller to another extension, press 4 to reach a fax handling menu, etc.). This takes too much time. Your ear can process information only one chunk at time, while your eyes scan a number of options in an instant.

Client-Server Switch Replacing The "Mainframe" PBX — Will It Ever Happen?

From a *TELECONNECT Magazine* article, this is a view of one possible way that the industry may develop. It is not yet clear if this will happen or if the PBX will hang in there in its present form, being controlled by PCs at the desktop or LAN level.

The traditional PBX can be likened to the computer mainframe, a central hub of processing power and information. It may someday be replaced with what has been called a *client-server switch.* It appears that the architecture of these switches will be the following:

1. Switches will be servers on local area networks or they will be located in other servers, depending upon the size

of the switch. There will be workgroup switches, departmental switches, floor switches, building switches, etc. These will be joined by a hierarchical design and by many different network interfaces.

2. Network or central office switches (outside of the office, run by telephone companies) will follow the same architecture, distributed in small boxes and joined by high-speed rings of cable. This will eventually replace the central office switches that can now take up a building the size of a city block.

The most distinguishing feature will be modularity in both hardware and software. This is made possible by two features not found in today's switches: (1) an operating system and (2) the PC's form with its inexpensive electronics and power supply.

Virtually all switches today are without a standard general purpose operating system. Since they were designed as computers dedicated to switching telephone calls, the need for more general purpose programming was not considered. At the time most switches were designed, computers were still expensive and not as powerful as they have become. Thus, the designer's perspective was that their switches would run faster skipping the overhead of an operating system and writing their code to a small "kernel." It worked, but resulted in a closed and inflexible architecture.

An open and public operating system will mean that people other than the manufacturer can program the switch, making it do things that the manufacturer never thought of or did not have time to address. The switch will do what the customer wants it to do. This is the core of the "open" telecommunications revolution.

The software that will sit on the operating system will be object-oriented, chunk-sized and modular.

Here is the situation which is driving the client/server switching development as it is today. The average PBX contains three to five million lines of software code. Adding a few new features requested by a customer requires 100,000 lines of software code, taking several months of a programmer's time. Next, "regression testing" takes place. The new code is tested in combination with all of the old code, simulating conditions of the telephone system under normal day-to-day use. This may take another three to six months or more.

This is further complicated because it is unlikely that the programmers writing the new code are the same programmers who wrote the original code. *(Code* refers to the sequence of instructions in software.)

Today's PBX manufacturers respond to this phenomenon by releasing new system features infrequently. When they do, a whole collection of new features is introduced, resulting in a very long time for testing before the new software can be released. Once released, these software upgrades can be costly to install and may still have "bugs" to be worked out with fixes called *patches.*

If a customer wants just one of the new features in the new software, he must still pay for the entire upgrade.

As switches incorporate more features and thus more software code as demanded by the customers, the chance of system failures and the time required for testing are both increasing exponentially.

While the manufacturers are being conservative in not offering too many new features too quickly, the switch users are experiencing the *programmability* of PCs and want the same capability for the telephone systems.

The movement of client/server telephony is toward *object-oriented software.* Each object is actually a chunk of software that is written, tested and can work with other chunks, building up a library of usable chunks. There will be layers of software. Some will be for classic features such as Least Cost Routing and Call Conferencing. Some may be applications generators. Some may be script languages. The concept is to build a huge library of software chunks, dipping into it as necessary without cluttering one feature with many others, as is the case with telephone systems now.

The switch factory of the future, whether it's making PBXs, key systems or central office switches, may make each system to the specific requirements of the customer who ordered it. The factory will assemble the network interface modules, the hardware modules and the software modules or chunks into a customized telecommunications solution. The switchmaker's skill where value is added will be in three areas:

1. The Network Interface – The switch manufacturer will make it possible to connect to T-1, E1 (European equivalent of T-1), SS7, etc. This is not trivial. There are dozens of interfaces to telecommunications networks. There are standards, but the tendency has been towards non-standard interfaces. For example, there are at least 20 varieties of ISDN Basic Rate Interface lines in North America, and even more internationally.

2. Adding "robustness" to the switch, meaning making it rugged, durable, redundant, and able to support areas which could be weak such as network interfaces and supplies.

3. Testing of products manufactured by others to work with the switch. The switch manufacturer will recommend the software modules, objects and peripheral hardware that will

work best with their system. It remains to be seen whether or not they will guarantee the performance of the products made by the other companies.

The benefits of the client/server switch architecture include the following:

1. It is customized to your individual needs. You pay for what you need, not what you get (as with today's PBXs, where you may never use many of the system features).
2. It can be truly different from what everyone else has and therefore provides a real competitive edge for your customers.
3. You can get the equipment and hardware you prefer since hardware, software and telephones will all work on most platforms. Even the PC interfaces can be tailored to provide that with which the individual is most comfortable.
4. New features can be added quickly.
5. The life cycle of switches will decline, approaching that of PC's and local area networks.

As the industry develops, time will tell how much of the above actually happens and how quickly.

Network As A Platform

Long distance networks and the services they provide have become commodities. It is becoming increasingly difficult to distinguish one from another. The next attempt at differentiation will be with the specialized network, the network as a platform.

Think of the network as a computer operating system on which others can build applications. The more operating systems sold, the more usage on the network.

It has been estimated that the cost of providing a digital circuit mile of telephone capacity has dropped by 98 percent in the past ten years. With the deployment of fiber optic cable throughout the carrier networks, theoretically there is no capacity limit. Additional telephone calls may be added to the network by adding electronics at either end.

The telephone industry has never before pursued the idea that a specialized network could offer its customers a competitive edge. The carriers have only cared about the conversations in progress from point A to point B, not about the business purposes for which the network was being used. The networks have become competitive tools and the long distance carriers are now realizing the potential value that outside developers may bring.

The carriers' biggest investments of the past decade have been in two areas. The first is the construction of a network separate from the one that is carrying all of the telephone traffic. That network is called *SS7, Signaling System 7.* It acts as a traffic cop for network traffic, assigning highways and routing customer traffic. SS7 is what makes the long distance carriers able to set up coast-to-coast calls in less than one second.

The second big investment has been in large billing and intelligence systems.

Two aspects of the open long distance networks which create the most exciting opportunities for developers are SS7 and the carrier's billing systems. Tying into SS7 means being able to control the movement of calls from the instant that they hit the network. That is more efficient than waiting until the calls hit your office and then sending them back to the network for rerouting. In days past, a call hunted its way through a hierarchy of switches across the country. With SS7, a call from New

York to Los Angeles is held in New York only until SS7 clears a path for it.

Tying into SS7 can do more than just determine when the carriers' switches are full. It can tell the network which customer service representative in your call center is able to take the next 800 call irrespective of that representative's location. It can ensure returned calls even though the line may be busy. The possibilities are endless.

The second opportunity for developers is the carrier's billing systems. There, the capability for customers to buy things over the telephone without using credit cards exists.

Here are a few examples of open network services:

1. You are sending a fax to someone 2,000 miles away. The fax line is busy. You send a message, "Please tell me when the line becomes free. Hold it open while I send my fax."
2. You advertise your product. "Call this 800 number to speak to your closest dealer." A customer calls. You have the calling number. You know where the customer is calling from and send him to your closest dealer. If the dealer does not answer, you send the customer to the next-closest dealer.
3. You carry a cellular telephone. When you travel to a new city, you punch in a few digits. When someone calls you, they dial your direct telephone number in your office but it does not ring. Instead, your cellular telephone thousands of miles away rings. The call finds you.

We have highlighted some of the emerging concepts and technologies that make up *Computer Telephony*. New products and services will continue to be released at a rapid pace. It's an exciting time for the computer and telecommunications industries.

There will be many clashes and hopefully, a lot of cooperation. This book is in the spirit of that cooperation.

For additional reading on Computer Telephony see 1001 Computer Telephony Tips, Secrets & Shortcuts and Client Server Computer Telephony,, both by Edwin Marguilies (order from 1-800-LIBRARY) and subscribe to "*Computer Telephony Magazine*."

Index

Symbols

A

B

C

D

N

O

U

V

W

Z